Markt- und Unternehmensentwicklung
Markets and Organisations

Change of institutions, technology and competition drives the interplay of markets and organisations. The scientific series 'Markets and Organisations' addresses a magnitude of related questions, presents theoretic and empirical findings and discusses related concepts and models.

Martin Dumbach

Establishing Corporate Innovation Communities

A Social Capital Perspective

Springer Gabler

Martin Dumbach
Innovation & Value Creation
Friedrich-Alexander-University
Nuremberg, Germany

Dissertation Friedrich-Alexander-University of Erlangen-Nuremberg, 2013

Erstreferent: Prof. Dr. Kathrin M. Möslein
Zweitreferent: Prof. Dr. Michael Amberg
Letzte Prüfung: 14.05.2013

ISBN 978-3-658-03694-2 ISBN 978-3-658-03695-9 (eBook)
DOI 10.1007/978-3-658-03695-9

The Deutsche Nationalbibliothek lists this publication in the Deutsche Nationalbibliografie; detailed bibliographic data are available in the Internet at http://dnb.d-nb.de.

Library of Congress Control Number: 2013951783

Springer Gabler

Printed on acid-free paper

Springer Gabler is a brand of Springer DE.
Springer DE is part of Springer Science+Business Media.
www.springer-gabler.de

Geleitwort

Die Innovationsfähigkeit von Organisationen gilt zunehmend als Erfolgsfaktor im Wettbewerb. In der Konsequenz haben sich in Wissenschaft und Wirtschaftspraxis zahlreiche Instrumente etabliert, die dem Management bei der Ausgestaltung und Unterstützung von Innovationsprozessen helfen. Neuere Ansätze fokussieren in diesem Zusammenhang verstärkt auf die Ausgestaltung von Innovationsprozessen über organisationale Grenzen hinweg. Dabei spielen sowohl Mitarbeiter der betreffenden Unternehmen aus unterschiedlichen Organisationseinheiten als auch externe Stakeholder, wie Kunden, Lieferanten oder Universitäten, eine zentrale Rolle. Gemeinschaften von unternehmensinternen und –externen Akteuren, die online oder offline in Interaktion treten und Innovationen treiben, gewinnen für Unternehmen zunehmend an Bedeutung. Sie werden gezielt angestoßen, von Unternehmen betrieben und gelten als wichtiger Hoffnungsträger und Treiber für Innovationsfähigkeit und Innovationserfolg von Unternehmen. Dabei kommunizieren die Mitglieder sowohl über IT-basierte Plattformen als auch in Face-to-Face-Settings, um gemeinsam Innovationsaufgaben zu lösen. Derartige unternehmensgetriebene Innovationsgemeinschaften werden beispielsweise von den Unternehmen Dell (www.ideastorm.com) oder LEGO (www.lego.cusoo.com) betrieben.

Innovation Communities sind im Zuge ihrer verstärkten Verbreitung und Bedeutung in der Unternehmenspraxis auch seit Jahren intensiver Untersuchungsgegenstand der Forschung. Viele Aspekte gelten heute als mehr oder weniger gesichert. Während jedoch in der Literatur beispielsweise weitgehende Einigkeit darüber besteht, dass Sozialkapital ein zentraler Erfolgsfaktor für Innovationsgemeinschaften ist und dessen Implikationen vielfach diskutiert sind, sind die Vorbedingungen seines Entstehens weit weniger klar. Diese Lücke zu schließen, ist das zentrale Anliegen der Dissertationsschrift von Martin Dumbach.

Die Forschungslücke ist von Martin Dumbach mit Weitblick identifiziert: Bedingungen für die Entstehung sozialen Kapitals zu schaffen, kann für die zukünftige Gestaltung von Innovationsgemeinschaften als zentraler Hebel betrachtet werden. In der Identifikation dieser Lücke liegt daher bereits die erste beeindruckende Forschungsleistung des Verfassers: Die Lücke liegt nicht auf der Hand. Ihre Identifikation ist nur vor dem Hintergrund einer überaus breiten und tiefen Literaturkenntnis möglich. Das Potential der Beantwortung der selbst gewählten Forschungsfrage („What are the antecedents of social capital in corporate innovation communities?") hingegen liegt auf der Hand: Wenn Sozialkapital entscheidender Erfolgsfaktor für die Gestaltung von Innovationsgemeinschaften ist, dieses sich aber nicht direkt gestalten lässt, dann gilt es die Bedingungen seines Entstehens zu beschreiben, verstehen, erklären und gestalten. Die Bearbeitung der überaus anspruchsvollen Forschungsfrage ist daher nicht nur wissenschaftlich höchst interessant, sondern darüber hinaus für die praktische Gestaltung von Innovationsgemeinschaften außerordentlich relevant. Sie gelingt dem Verfasser mit beeindruckender Eleganz. Die Arbeit ist höchst anspruchsvoll angelegt sowie theoretisch wie empirisch überaus fundiert umgesetzt.

Das Buch verdient daher eine rasche Verbreitung in der Fachwelt. Es sei besonders all jenen ans Herz gelegt, die mit der strategischen Führung von Innovation in Organisationen befasst sind. Allen Lesern wünsche ich eine erkenntnisreiche Lektüre, vor allem aber Erfolg beim wissenschaftlichen Fortentwickeln oder wirtschaftlichen Umsetzen der erkenntnisreichen Einblicke in die Welt von Innovation Communities aus der theoretischen Perspektive des sozialen Kapitals.

Prof. Dr. Kathrin M. Möslein

Vorwort

Die Relevanz von informellen Netzwerken für unternehmerischen Erfolg wurde mir bereits während meiner ersten beruflichen Erfahrungen bewusst. Ob in meinem Ausbildungsbetrieb, während Praktika oder Werkstudententätigkeiten: Aufgaben, Herausforderungen und Probleme wurden zu einem nicht unerheblichen Ausmaß durch informelle Kommunikation außerhalb geregelter Prozesse gelöst. Dabei habe ich mich regelmäßig gefragt, welche Faktoren die Entstehung funktionierender Netzwerke bestimmen bzw. welche Einflussmöglichkeiten das Management in diesem Zusammenhang hat. Im Rahmen dieser Dissertation hatte ich die Möglichkeit, die Entstehung von informellen Communities im Kontext der Innovationsfähigkeit von Unternehmen wissenschaftlich zu untersuchen und so erste Antworten auf diese Fragen zu finden. Die Arbeit hat mir großen Spaß bereitet. Daher möchte ich mich bei allen bedanken, die mich in dieser Zeit unterstützt haben.

Allen voran möchte ich meiner Doktormutter Prof. Dr. Kathrin M. Möslein für die Möglichkeit an ihrem Lehrstuhl zu forschen und zu lehren danken. In meiner Zeit am Lehrstuhl habe ich sehr viel gelernt. Die Unterstützung und die Verbesserungsvorschläge meiner Professorin haben einen großen Anteil daran. Vielen Dank für das Vertrauen und den Optimismus meiner Arbeit gegenüber. Prof. Dr. Michael Amberg danke ich für die Übernahme des Zweitgutachtens.

Zudem möchte ich allen Mitarbeitern des Lehrstuhls Wirtschaftsinformatik I – Innovation & Wertschöpfung an der Friedrich-Alexander-Universität Erlangen-Nürnberg für die gute Zusammenarbeit danken. Hervorheben möchte ich Dr. Frank Danzinger, der als Postdoc diese Arbeit mitbetreut hat. Ohne seine Ratschläge wäre diese Arbeit nicht möglich gewesen. Anke Wendelken danke ich für das unkomplizierte Teamwork im Projekt WiIPOD[1]. Zudem danke ich Matthias Rass für hilfreiche Diskussionen im Rahmen unserer Doktorarbeiten. Besonders für die

---

[1] Das Projekt WiIPOD (Wertschätzungsnetzwerke als integrierte Innovationsinstrumente der Personal- und Organisationsentwicklung im Demografischen Wandel) wird vom Bundesministerium für Bildung und Forschung sowie dem Europäischen Sozialfonds unter den Förderkennzeichen 01HH11055-57 gefördert.

Unterstützung in meinen ersten Monaten am Lehrstuhl möchte ich mich bei Dr. Uta Renken bedanken.

Danken möchte ich auch allen Mitarbeitern des Projektes WiIPOD, die am Institut für Soziologie der Friedrich-Alexander-Universität sowie am Lehrstuhl für Persönlichkeitspsychologie und Psychologische Diagnostik der Otto-Friedrich-Universität Bamberg beschäftigt sind oder waren: Christian Sandig, Claudia Schmitt, Ronald Staples und Sabine Wabro. Die Zusammenarbeit mit Euch war mir eine Freude.

Weiterhin danke ich Prof. Dr. Andreas König, dem Betreuer meiner Diplomarbeit, dafür, dass er mir die Grundlagen wissenschaftlichen Arbeitens beigebracht hat, sowie Prof. Dr. Rainer Trinczek für seine Kommentare zu einer früheren Version dieser Dissertation.

Ganz besonderer Dank geht an meine Familie und meine Freunde für die Unterstützung über viele Jahre hinweg. Meinen Freunden Björn Gottschall und Florian Jung danke ich für viele lustige Stunden im In- und Ausland. Meinen Eltern Hannelore und Josef sowie meinen Schwestern Evelin und Carolin, meiner Nichte Johanna und meinem Neffen Moritz, danke ich, dass sie immer für mich da sind und waren.

Die umfangreichste Unterstützung im Rahmen dieser Dissertation hat Hanna Plieth geleistet. Vielen Dank für das Verständnis für die zahlreichen Stunden, die diese Arbeit auch am Wochenende gekostet hat, sowie für die hilfreichen Anregungen und Verbesserungsvorschläge.

Martin Dumbach

Overview of contents

List of figures XVI
List of tables XVII
List of abbreviations XIX

I Introduction: *What is this dissertation about?* 1
1 Innovation communities and social capital theory 8
2 The research gap: Limited knowledge on antecedents of social capital 11
3 Overview of studies 12

II Theoretical background: *Open innovation, communities, and social capital* 17
1 Structure of part II 19
2 The paradigm of open innovation 21
3 The central characteristics of corporate innovation communities 27
4 The relevance of social capital in innovation processes 33
5 Derivation of research gap and research question 52

III Antecedents of social capital in corporate innovation communities on three levels: *A case study approach* 55
1 Structure of part III 57
2 General methodological issues 60
3 Case selection 63
4 Data collection and analysis 70
5 Findings on antecedents of corporate innovation community social capital 79
6 Discussion 123

IV Individual level antecedents of social capital in corporate innovation communities: *A SEM approach* 129
1 Structure of part IV 131
2 Theoretical foundations 136

3 Model development 161
4 Empirical field 175
5 Data analysis approach 181
6 Measures and measurement validity 190
7 Hypotheses testing 208
8 Discussion 212

V Discussion and conclusion: *Reflections on research presented in this dissertation* 217

1 Summary of studies 219
2 Theoretical implications 223
3 Managerial implications 227
4 Limitations and avenues for future research 236
5 Conclusion 239

References 240

Annexes 265

Table of contents

List of figures ..... **XVI**
List of tables ..... **XVII**
List of abbreviations ..... **XIX**

I Introduction: *What is this dissertation about?* 1

1 Innovation communities and social capital theory **8**
2 The research gap: Limited knowledge on antecedents of social capital **11**
3 Overview of studies **12**

II Theoretical background: *Open innovation, communities, and social capital* 17

1 Structure of part II **19**
2 The paradigm of open innovation **21**
3 The central characteristics of corporate innovation communities **27**
4 The relevance of social capital in innovation processes **33**
4.1 Social capital: Definition and relevance 34
4.2 The three dimensions of social capital 39
4.2.1 The structural dimension of social capital 40
4.2.2 The relational dimension of social capital 43
4.2.3 The cognitive dimension of social capital 44
4.3 Social capital – a multilevel concept 45
4.3.1 Individual level studies 47
4.3.2 Group level studies 48
4.3.3 Organizational level studies 49
4.3.4 Cross-level studies 50
5 Derivation of research gap and research question **52**

III Antecedents of social capital in corporate innovation communities on three levels: *A case study approach*.......... 55

1 **Structure of part III**.......... 57
2 **General methodological issues**.......... 60
3 **Case selection**.......... 63
3.1 Pilot Case A: Temporary employment.......... 64
3.2 Four follow up cases.......... 65
3.2.1 Case B: Automobile.......... 65
3.2.2 Case C: Toy manufacturing.......... 66
3.2.3 Case D: Automation.......... 67
3.2.4 Case E: Conglomerate.......... 68
4 **Data collection and analysis**.......... 70
4.1 Data collection: Case A.......... 70
4.2 Data collection: Case B-E.......... 73
4.2.1 Data collection: Case B.......... 73
4.2.2 Data collection: Case C-E.......... 74
4.3 Data analysis.......... 77
5 **Findings on antecedents of corporate innovation community social capital**.......... 79
5.1 The self-reinforcing nature of innovation community social capital.......... 81
5.2 Individual level antecedents of innovation community social capital.......... 89
5.2.1 Personality traits.......... 90
5.2.2 Intrinsic and extrinsic motivations.......... 96
5.3 Organizational level antecedents of innovation community social capital.......... 99
5.3.1 Corporate culture.......... 100
5.3.2 Formal and informal organization.......... 104
5.4 Community level antecedents of innovation community social capital.......... 110
5.4.1 Creative task design.......... 110
5.4.2 Community leadership.......... 112
5.4.3 Workshop and platform design.......... 115
6 **Discussion**.......... 123
6.1 Community social capital as a multidimensional phenomenon.......... 123
6.2 Community social capital antecedents on multiple levels.......... 124
6.3 Differences to extant studies on antecedents of social capital.......... 125
6.4 Limitations and avenues for future research.......... 127

IV Individual level antecedents of social capital in corporate innovation communities: *A SEM approach*................ 129

1 Structure of part IV 131

2 Theoretical foundations 136

2.1 Personality traits 136

2.1.1 The Five Factor Model of personality 138

2.1.1.1 Openness to experience 140

2.1.1.2 Agreeableness 142

2.1.1.3 Emotional stability 145

2.1.1.4 Introversion 147

2.1.1.5 Conscientiousness 149

2.2 Motivations of innovation community members 151

2.2.1 Extrinsic motivation 152

2.2.1.1 Usage of better products and services 153

2.2.1.2 Reputation and career perspectives 153

2.2.1.3 Financial rewards 154

2.2.1.4 Learning 155

2.2.2 Intrinsic motivation 157

2.2.2.1 Fun and enjoyment 157

2.2.2.2 Identification with the community 159

2.2.2.3 Increased knowledge self-efficacy 160

2.2.2.4 Positive emotions related to helping 160

3 Model development 161

3.1 Self-reinforcing relationships between structural, cognitive, and relational social capital in corporate innovation communities 163

3.1.1 Network size and shared visions, tie strength as well as trust 164

3.1.2 Shared visions and trust as well as tie strength 166

3.1.3 Tie strength and trust 167

3.2 The "Big 5" personality traits as antecedents of corporate innovation community social capital 168

3.2.1 Openness to experience and innovation community social capital 168

3.2.2 Agreeableness and innovation community social capital 169

3.2.3 Emotional stability and innovation community social capital 169

3.2.4 Introversion and innovation community social capital 170

3.2.5 Conscientiousness and innovation community social capital 171

3.3 Extrinsic and intrinsic motivations as antecedents of corporate innovation community social capital 172

3.3.1 Extrinsic motivation and innovation community social capital 173

3.3.2 Intrinsic motivation and innovation community social capital 173
4 Empirical field 175
4.1 WIN Innovation Community 175
4.2 Data collection approach 178
5 Data analysis approach 181
5.1 SEM and its use in this study 181
5.1.1 Exploratory factor analysis 183
5.1.2 Confirmatory factor analysis 184
5.1.3 Evaluation of model fit 185
6 Measures and measurement validity 190
6.1 Measurement of the FFM 190
6.2 Measurement of extrinsic and intrinsic motivations 191
6.3 Measurement of social capital 192
6.3.1 Measurement of the structural dimension of social capital 192
6.3.2 Measurement of the cognitive dimension of social capital 194
6.3.3 Measurement of the relational dimension of social capital 194
6.3.3.1 Trust 194
6.3.3.2 Tie strength 195
6.4 Measurement model validity 195
6.4.1 EFA based on the whole data set 197
6.4.2 EFA, CFA, and model fit on construct level 199
6.4.2.1 Validity of the measurement of the FFM 199
6.4.2.2 Validity of the measurement of intrinsic and extrinsic motivations 203
6.4.2.3 Validity of the measurement of social capital 204
7 Hypotheses testing 208
8 Discussion 212
8.1 Theoretical implications 214
8.2 Limitations and avenues for future research 215

V Discussion and conclusion: *Reflections on research presented in this dissertation* **217**
1 Summary of studies **219**
1.1 Summary of the study presented in part III 219
1.2 Summary of the study presented in part IV 221
2 Theoretical implications **223**
3 Managerial implications **227**
4 Limitations and avenues for future research **236**
5 Conclusion **239**

References **240**

Annexes **265**
Annex 1: Related publications and data collection **266**
Annex 2: German wording of items used in part IV **268**
Annex 3: Evaluation of the measurement of trust in part IV **271**
Annex 4: Covariance matrix of items used in part IV **273**

List of figures

Figure 1: Practical examples of innovation communities 6
Figure 2: Definition of corporate innovation communities and the relevance of networks 9
Figure 3: Research gap and research question 11
Figure 4: Summarized findings of the case study approach presented in part III 12
Figure 5: Summarized findings of the SEM approach presented in part IV 14
Figure 6: Open innovation and its relevance in innovation communities 26
Figure 7: Definition of social capital and its relevance in innovation communities 38
Figure 8: The three dimensions of social capital 40
Figure 9: The relevance of the multilevel nature of social capital 46
Figure 10: Structure of part III based on Eisenhardt (1989) 59
Figure 11: Case study methodology and characteristics of the present study 62
Figure 12: Advantages of single versus multiple case studies and research design 64
Figure 13: Structure of the proposed framework 80
Figure 14: Organizational environment and community social capital 81
Figure 15: Interrelated dimensions of social capital in corporate innovation communities 89
Figure 16: Individual level antecedents of social capital 99
Figure 17: Organizational level antecedents of social capital 109
Figure 18: Community level antecedents of social capital 121
Figure 19: Structure of part IV based on Hair et al. (2010) 135
Figure 20: Intrinsic and extrinsic motivations of innovation community members 152
Figure 21: Hypothesized model 162
Figure 22: Conceptualization of social capital 164
Figure 23: Steps to ensure successful data collection 178
Figure 24: Three step approach to examine validity of measurement and structural model .. 183
Figure 25: Hypothesized model and maximum likelihood parameter estimates 211
Figure 26: Summary of research findings 222
Figure 27: Summary of managerial implications 228

List of tables

Table 1: Central characteristics of corporate innovation communities ... 28
Table 2: Overview of cases ... 69
Table 3: Overview of interviews and other data sources ... 77
Table 4: The self-reinforcing nature of social capital ... 83
Table 5: The effect of structural on cognitive social capital ... 84
Table 6: The effect of structural on relational social capital ... 85
Table 7: The effect of cognitive on relational social capital ... 85
Table 8: The effect of cognitive on structural social capital ... 86
Table 9: The effect of relational on structural social capital ... 87
Table 10: The self-reinforcing destruction of social capital ... 88
Table 11: The role of openness to experience ... 91
Table 12: The role of agreeableness ... 92
Table 13: The role of emotional stability ... 93
Table 14: The role of extraversion ... 94
Table 15: The role of conscientiousness ... 95
Table 16: The role of intrinsic motivation ... 97
Table 17: The role of extrinsic motivation ... 98
Table 18: The role of collaborative corporate culture ... 101
Table 19: The role of innovative corporate culture ... 103
Table 20: The role of formal organization ... 105
Table 21: The role of informal organization ... 106
Table 22: The role of social supporting activities of organizations ... 107
Table 23: The role of technical supporting activities of organizations ... 108
Table 24: The role of creative task design ... 112
Table 25: The role of community leadership ... 114
Table 26: The role of workshop design ... 117
Table 27: The role of platform design ... 119
Table 28: Multilevel antecedents of corporate innovation community social capital ... 122
Table 29: The Five Factor Model of personality and relevant findings ... 140
Table 30: Usage of interaction channels in the WIN Innovation Community ... 176

Table 31: Overview of indices 189
Table 32: Factor loadings of EFA based on the Kaiser criterium 198
Table 33: Measurement validity of the construct "openness to experience" 200
Table 34: Measurement validity of the construct "agreeableness" 200
Table 35: Measurement validity of the construct "emotional stability" 201
Table 36: Measurement validity of the construct "introversion" 202
Table 37: Measurement validity of the construct "conscientiousness" 202
Table 38: Measurement validity of the construct "intrinsic motivation" 203
Table 39: Measurement validity of the construct "extrinsic motivation" 204
Table 40: Measurement validity of the construct "trust" 205
Table 41: Measurement validity of the construct "tie strength" 206
Table 42: Measurement validity of the construct "shared visions and languages" 207
Table 43: Change in χ^2*-tests of alternative models* 210
Table 44: German wording of items (I) 268
Table 45: German wording of items (II) 269
Table 46: German wording of items (III) 270
Table 47: Measurement validity of the alternative measurement of trust 272
Table 48: Covariance matrix (I) 273
Table 49: Covariance matrix (II) 274
Table 50: Covariance matrix (III) 275

List of abbreviations

AGFI	Adjusted Goodness of Fit Index
AVE	Average variance extracted
CBSEM	Covariance-based structural equations modeling
CEO	Chief executive officer
CFA	Confirmatory factor analysis
CFI	Comparative Fit Index
df	Degrees of freedom
EFA	Exploratory factor analysis
et al.	et alia (lat.); and others
e.g.	exempli gratia (lat.); for example
FAU	Friedrich-Alexander-University of Erlangen-Nuremberg
FFM	Five Factor Model
F-L-ratio	Fornell-Larcker-ratio
GFI	Goodness of Fit Index
HR	Human resources
i.e.	id est (lat.); that means
IT	Information technology
ICT	Information and communication technology
MSA	Measure of sampling adequacy
MSLQ	Motivated Strategies for Learning Questionaire
OEM	Original equipment manufacturer
OSS	Open source software
PC	Personal computer
RMSEA	Root Mean Square Error of Approximation
R&D	Research & development
SEM	Structural equations modeling
SRMR	Standardized Root Mean Square Residual
URL	Uniform resource locator
U.S.	United States

I

Introduction:

What is this dissertation about?

The ability to innovate is central for sustainable firm performance. More than seven decades ago, Schumpeter (1942) introduced the notion of creative destruction and argued that competitive advantages of companies are constantly in danger because competitors try to attack the market position of their rivals. Hence, organizations need to introduce new products, services or processes continually in order to keep ahead of the market. Since then, scholars have found empirical support for Schumpeter's (1939, 1942) arguments (Wiggins & Ruefli, 2005) and have discussed a number of environmental changes that reinforce the central role of innovation. For instance, Droge, Calantone, and Harmancioglu (2008) suggest that faster technological developments, rising research and development (R&D) costs, and shorter product life cycles amplify the importance of new product development. In addition, Lawson and Samson (2001) propose that the knowledge economy as well as intense global competition have strengthened the relevance of innovative performance. As another example, Schreyögg and Sydow (2010) add freedom of choice of individuals as well as rising technological complexity to the list of intensifiers of the central role of innovation.

Similarly, Porter states that "[c]ompanies achieve competitive advantage through acts of innovation. They approach innovation in its broadest sense, including both new technologies and new ways of doing things" (1991: 136). There are numerous innovations that have helped companies to establish a prosperous market position. In this context, companies operating in high-tech sectors such as the computer industry, for instance IBM or Intel, are often mentioned as illustrative examples (e.g., Chesbrough, 2003a). However, innovation plays also a pivotal role in low-tech industries. For instance, the Gillette Company introduces new and improved shaving systems (e.g., Sensor or Mach III) continuously and defends its market position against competitors successfully (Kelley, 2006). The company Dell offers an example for innovative activities of companies that address the way how value creation is organized (i.e., the "new ways of doing things" (Porter, 1991:136) aspect of innovation mentioned in the definition above). Dell changed the computer industry dramatically by skipping one step in the value chain and selling its computers directly to customers instead of relying on retail businesses. As a result, Dell is famous as a "lean distributor of computers" (Bhaskaran & Krishnan, 2009: 1153).

The list of innovations that have influenced success and defeat in competitions among companies is huge as well as diverse. At the same time, there is general agreement that traditional ways of organization and management which focus on efficiency are not suitable to support innovation processes in organizations. As Judge, Fryxell, and Dooley put it: "Due to the ambiguous, unique, and inherently uncertain process of industrial innovation, research and development units must be managed quite differently from other work units where the work is more routine" (1997: 72). Sawhney and Prandelli (2000) take the same line and suggest that the organization of innovation processes has to allow for flexible knowledge exchange between relevant stakeholders. In reaction, a number of scholars, including Judge et al. (1997) and Sawhney and Prandelli (2000), have argued that companies can make use of the concept of communities in order to support innovation processes[2]. Within innovation communities, individuals exchange knowledge and ideas freely and in an informal manner (Fleming & Waguespack, 2007). As such, the concept of communities fits properly to innovation processes (Brown & Duguid, 1991).

The development of the 3M Post-it Notes, an often referred to example in the literature (e.g., Garud, Dunbar, & Bartel, 2011), highlights the relevance of flexible and informal knowledge exchange in innovation processes. This is how 3M describes the process on its website[3]: 'Art Fry, a chemist and researcher at 3M, was a member of the local church choir and regularly annoyed because his bookmarks dropped out of his music book. Luckily, he talked to his colleague Dr. Spencer Silver about this problem, who had just invented a glue that was sticking very lously. Art Fry started to brush his bookmarks with this newly developed glue and helped 3M to find a useful application for its lously sticking glue, a concept today known under the brand Post-it Notes'. Obviously, informal knowledge exchange played a pivotal role within this innovation process.

In innovation communities, employees and external stakeholders engage in informal interaction outside hierarchical structures. The concept originates from open source software (OSS), where many community members contribute voluntarily and

[2] Of course, there are other approaches such as stage-gate processes (Cooper, 2001) to solve the innovation challenge as well. These approaches are not discussed within this dissertation.

[3] http://solutions.3mdeutschland.de/wps/portal/3M/de_DE/Post-Its/Post-It/Solutions/History (accessed December 2012).

informally to software code that is made available to the community afterwards (Lerner & Tirole, 2002). Today, innovation communities have also been analyzed in other industries (e.g., sports (von Hippel, 2001)) and a number of researchers have discussed how companies can make use of the concept in the organization and management of innovation processes (Nalebuff & Ayres, 2004; Pisano & Verganti, 2008).

Figure 1 shows two popular examples of innovation communities and provides short descriptions, URLs and screenshots of online community platforms[4]. The first community is supported by Dell and is coined IdeaStrom. The IdeaStorm community has been analyzed in many studies (Bayus, 2012; di Gangi & Wasko, 2009; di Gangi, Wasko, & Hooker, 2010) and allows Dell to interact with its customer base in order to grasp innovative ideas. Dell provides the community platform, which allows members to submit ideas to the community as well as to comment and promote the ideas of others. As depicted on the screenshot, members submitted over 18,000 ideas to the community since 2007 and the company has implemented more than 520 ideas. The second community depicted in Figure 1 is supported by LEGO and is named CUUSOO. As is the case with Dell, many researchers have referred to LEGO's communities (Frey & Lüthje, 2011; Lettl, Rost, & von Wartburg, 2009) where the company interacts with external stakeholders in order to grasp innovative product ideas. The CUUSOO platform allows members to evaluate and discuss ideas of other members as well as to submit own ideas. As the description of the community in Figure 1 shows, successful community members capture 1% of the revenue generated with their ideas. Similar community platforms like the one of Dell and LEGO are also used to support innovative exchanges within the boundaries of companies (i.e., among employees), for example within the IBM innovation jam initiative (Bjelland & Wood, 2008).

[4] Community platforms have been accessed December 2012 to gather the information depicted in Figure 1.

1 Dell IdeaStorm

Description on website

"IdeaStorm was created to give a direct voice to our customers and an avenue to have online "brainstorm" sessions to allow you the customer to share ideas and collaborate with one another and Dell. Our goal through IdeaStorm is to hear what new products or services you'd like to see Dell develop. We hope this site fosters a candid and robust conversation about your ideas."

Screenshot

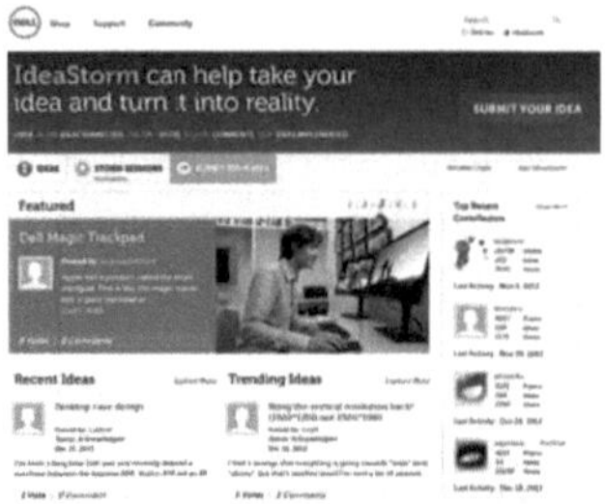

URL: http://www.ideastorm.com

2 LEGO CUUSOO

Description on website

"1. Have an Idea? Create a Project, share your product concept, and see what other people think.
2. Get 10,000 Supporters. Projects with 10,000 supporters are reviewed quarterly by LEGO for a chance to become an official LEGO product.
3. Receive Royalty . If your project passes review and is chosen for production, you will receive 1% of the total net sales of the product."

Screenshot

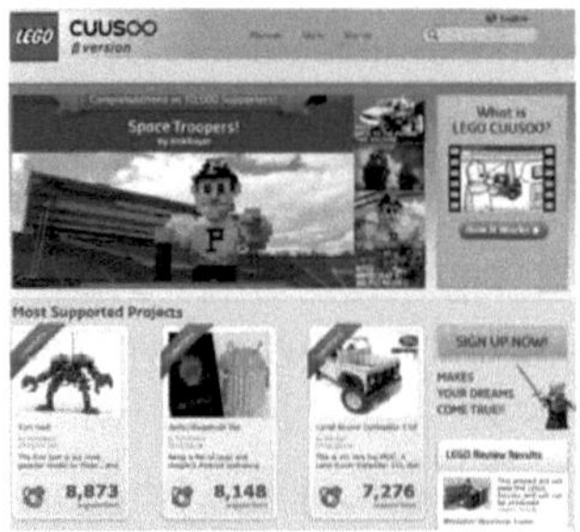

URL: http://lego.cuusoo.com

Figure 1: Practical examples of innovation communities

Because innovation communities are different to traditional ways of organizing, most importantly because hierarchical structures are absent (Fleming & Waguespack, 2007), social relationships and networks between community members are important. In essence, these networks determine the extent to which knowledge and ideas are exchanged between community members (Dahlander & Frederiksen, 2012; McLure Wasko & Faraj, 2005). Related to this informal and hierarchy-free character, many people intuitively assume that innovation community networks are hard to manage and researchers have provided theoretical rationales for this assumption (e.g., Balkundi & Kilduff, 2005; Uzzi, 1997). As a result, those who are responsible to manage community networks fall short with regard to recommendations how to run innovation communities successfully.

The research presented in this dissertation offers insights for both researchers and practitioners interested in innovation community management. From a theoretical point of view, the studies presented contribute to the literature on social relationships within innovation communities by taking a social capital perspective (Nahapiet & Ghoshal, 1998). The research presented adds to the understanding of antecedents of social capital in community settings, a facet that has not received enough attention in the literature so far (Payne, Moore, Griffis, & Autry, 2011).

From a practical point of view, the studies presented are helpful as they disclose the complexity of community management and highlight that community management is not the task of a single position, but involves both operational activities of innovation and human resource (HR) management as well as strategic decisions of executives. In contrast to the wide-spread believe that community networks are hard to influence, the framework presented shows that levers to support community networks are numerous. These levers concern strategic aspects with regard to organizational structures and corporate culture as well as operational aspects, for instance recruiting activities and incentives for community participation. In addition, community management is able to affect networks of community members via design elements of community workshops and platforms. The remainder of this introduction provides short insights into relevant literature streams, presents the research gap approached and gives an overview on the empirical studies within this dissertation.

1 Innovation communities and social capital theory

Recently, the role of communities that engage in innovation activities has received increasing attention in the academic literature (Bansemir, Neyer, & Möslein, 2012; Franke & Shah, 2003). For instance, Füller, Matzler, and Hoppe highlight the role of brand communities in the context of innovation and state: "Members of brand communities are considered an especially valuable source of innovation because they are passionate about the brand and experienced with its products [...] and because they support each other in solving problems and in generating new product ideas" (2008: 609). As another example, di Gangi and Wasko argue: "[U]ser innovation communities [...] are becoming increasingly influential in an organization's decision-making process regarding the adoption of new innovations" (2009: 303).

While many researchers have studied innovation communities, the understanding of an innovation community differs significantly among authors. For example, Füller and colleagues (2008) study innovative activities of a brand community of car enthusiasts who meet physically, while other scholars investigate innovation communities that interact online (Bansemir et al., 2012; Jeppesen & Laursen, 2009; Lakhani & von Hippel, 2003). At the same time, scholars have studied innovative communities inside the boundaries of organizations (Brown & Duguid, 1991), outside the boundaries of organizations (Franke & Shah, 2003) as well as innovative communities consisting of both employees of a company and external members (Bjelland & Wood, 2008; Dahlander & Wallin, 2006). Because of the multifaceted nature of the concept of innovation communities, West and Lakhani emphasize the importance that scholars "explicitly articulate the theoretical and phenomenological boundaries of their use of the term" (2008: 224). In this dissertation, I focus on communities that receive corporate support and name these communities corporate innovation communities. I define corporate innovation communities as *heterogeneous* associations of actors who *voluntarily* perform innovative activities in a *joint effort* in both *online* (e.g., IT-based innovation platforms) as well as *offline* (e.g., workshops) environments. A *company supports* the activities of community members who may or *may* not *be employed* by the focal organization. As such, my definition, contrary to the definition of West and Lakhani (2008), acknowledges innovative activities of communities inside as well as outside the boundaries of organizations.

Innovation communities operate within environments characterized by boundary crossing knowledge and idea flows and a lack of "pecuniary incentives, hierarchical authority, and formal structure" (Fleming & Waguespack, 2007: 166). The strength of innovation processes within such communities lies in the ability of community members to exchange knowledge and ideas informally and in a flexible way (Brown & Duguid, 1991; Franke & Shah, 2003; Muhdi & Boutellier, 2011). In such settings, social relationships between community members are a crucial success factor as these relationships determine access to knowledge and ideas of others and, as a consequence, affect innovativeness of community members (Dahlander & Frederiksen, 2012). For example, McLure Wasko and Faraj (2005) find that centrality in networks of innovative communities is positively associated with volume as well as helpfulness of contributions to the community. Figure 2 summarizes the need for an explicit definition of innovation communities, the understanding of corporate innovation communities in this dissertation as well as the resulting relevance of social networks in such settings.

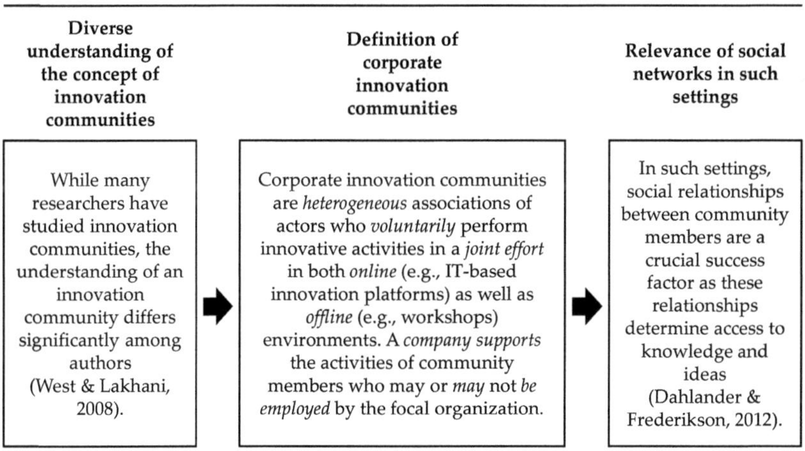

Figure 2: Definition of corporate innovation communities and the relevance of networks

The concept of social capital provides a theoretical framework that enables researchers to examine networks of relationships between individuals, groups and organizations (Nahapiet & Ghoshal, 1998; Payne et al., 2011). In this dissertation, I take a social capital perspective on corporate innovation communities. Scholars have often relied on social capital theory in order to examine how social relationships influence innovative outcomes (Tsai & Ghoshal, 1998) and new product development (Hansen, 1999). There is consensus among researchers that social capital affects innovation processes positively as it fosters knowledge exchange (Cross & Sproull, 2004; Reagans & McEvily, 2003) as well as creativity (Lingo & O'Mahony, 2010; Perry-Smith & Shalley, 2003). Hence, I suggest that the literature on social capital provides a suitable theoretical background to investigate social relationships in corporate innovation communities.

In their seminal article, Nahapiet and Ghoshal define social capital as "the sum of the actual and potential resources embedded within, available through, and derived from the network of relationships possessed by an individual or social unit" (1998: 243). Furthermore, Nahapiet and Ghoshal (1998) distinguish between three dimensions of social capital, i.e., (1) the structural dimension, (2) the relational dimension, and (3) the cognitive dimension. First, the structural dimension refers to the "structural position" (Koka & Prescott, 2002: 796) of the focal actor in a network. Second, the relational dimension refers to the quality of relationships (e.g., trust or frequency and closeness of interaction) within a network (Moran, 2005). Finally, the cognitive dimension of social capital concerns shared visions (Leana & Pil, 2006) and languages (Bolino, Turnley, & Bloodgood, 2002) among network members. It is important to emphasize that social capital exists at multiple levels (Payne et al., 2011). Researchers have repeatedly highlighted the multilevel characteristic of social capital (Moliterno & Mahony, 2011) and have analyzed network relationships on individual, group, and organizational levels (Brass, Galaskiewicz, Greve, & Tsai, 2004). This complex nature of social capital is important in the context of the research question approached in this dissertation, which I describe in the next chapter.

2 The research gap: Limited knowledge on antecedents of social capital

Though extant studies on innovative activities and social capital of communities are helpful, one particular outstanding issue remains. As already mentioned, there is consensus in the literature that social capital is a crucial success factor in innovation communities (Dahlander & Frederiksen, 2012; Fleming & Waguespack, 2007). However, relatively little is known with respect to the antecedents of social capital (Payne et al., 2011). Those scholars who examine antecedents of social capital focus on single dimensions (e.g., Brass et al., 2004) or on single levels of analysis (e.g., Bolino et al., 2002). As an example, Dahlander and Wallin (2006) study how the fact that an individual is sponsored by a firm influences the position of the focal individual in the community network. Thus, Dahlander and Wallin (2006) focus on the individual level as well as the structural dimension of social capital. The goal of this dissertation is to draw a more holistic picture of antecedents of innovation community social capital (see Figure 3). To address the research gap outlined in this paragraph, I consider multiple dimensions as well as multiple levels of social capital and ask: *What are antecedents of social capital in corporate innovation communities?*

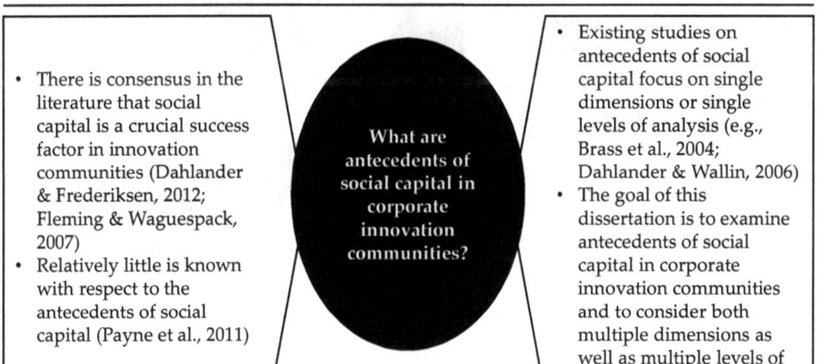

Figure 3: Research gap and research question

3 Overview of studies

In order to answer the research question raised in the previous chapter, I present two interrelated studies in this dissertation. The first study is a multiple case study as proposed by Eisenhardt (1989). Qualitative data has been collected in five corporate innovation communities supported by different organizations. As recommended by Bogers (2011), the supporting organizations examined are heterogeneous in terms of size and industry characteristics. While my primary data source was semi-structured interviews, I triangulated interview data with a number of other sources such as community workshop observations or analysis of interaction on online platforms. In total, I analyzed 72 interviews. In addition, I observed 14 community workshops and analyzed interaction on three online community platforms. The findings of the multiple case study are summarized in Figure 4.

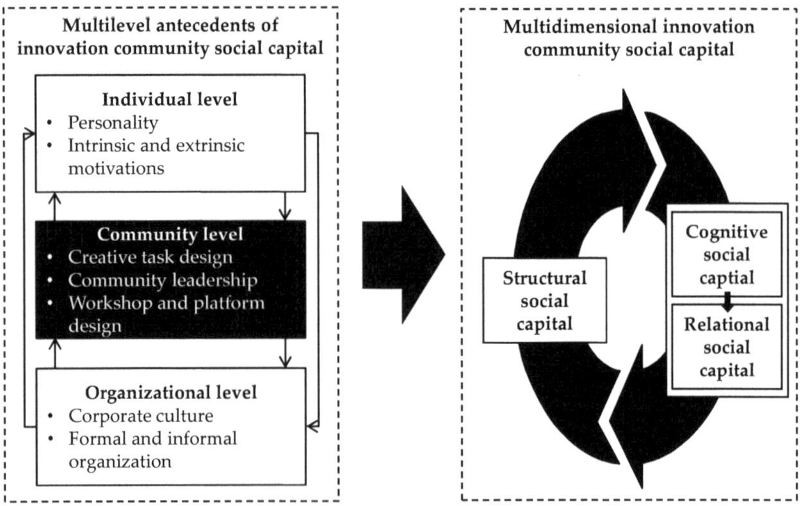

Figure 4: Summarized findings of the case study approach presented in part III

Figure 4 depicts two important aspects. First, it suggests that dimensions of social capital are interrelated and reinforce themselves. That is, the structural dimension of social capital (i.e., interaction ties between community members) is an antecedent of the cognitive (i.e., shared visions and languages) as well as the relational dimension (i.e., closeness and trust) of social capital in corporate innovation communities. In other words, interaction with others creates opportunities to develop stronger relationships, trust, as well as a shared vision and language. Furthermore, the cognitive dimension antecedes the relational dimension. Shared visions and languages enable close and trusted exchange between community members. On the other hand, shared visions and trust between community members lead to further interaction ties within the community. Second, as the framework in Figure 4 suggests, antecedents on three interrelated levels affect the self-reinforcing cycle of social capital. Data indicates that, on the individual level, personality and motivations of community members affect social capital. On the organizational level, empirical observations suggest that corporate culture as well as formal and informal organization of the supporting company provide environmental aspects of community activities that affect social capital. Finally, on the community level, data suggests that appropriate task design, community leadership as well as workshop and platform design affect community social capital. As social capital represents a crucial success factor in innovation communities (Dahlander & Frederiksen, 2012; McLure Wasko & Faraj, 2005), the dynamics depicted in Figure 4 offer interesting insights. A core contribution of the framework presented is the consideration of multiple dimensions as well as multiple levels of social capital, a facet that has not received enough attention in studies on antecedents of social capital so far.

In the second study, I take a quantitative, structural equations modeling (SEM) approach (Diamantopoulos & Siguaw, 2000; Gefen, Rigdon, & Straub, 2011). The empirical field is an innovation community composed of students of the Friedrich-Alexander-University of Erlangen-Nuremberg (FAU) (n=993). Within this study, I test aspects situated at the individual level of corporate innovation community social capital. In more detail, I examine relationships between personality as well as extrinsic and intrinsic motivations and interrelated dimensions of community social capital. Exogeneous constructs of the model are personality factors of the Five Factor Model (FFM) as well as extrinsic and intrinsic motivations. Endogeneous constructs are network size (structural social capital), shared visions and languages (cognitive social capital) as well as tie strength and trust (relational social capital). Figure 5 summarizes my findings. I find support for the hypothesized interrelations between different dimensions of social capital. Furthermore, I find that aspects of personality affect structural as well as cognitive dimensions of social capital and that intrinsic motivation affects the cognitive dimension of social capital. Data did not confirm a significant relationship between extrinsic motivation and social capital in corporate innovation communities. Compared to the study presented in part III, the study presented in part IV takes a closer focus and contributes to the literature by empirically testing the significance of antecedents of social capital in corporate community settings on the individual level.

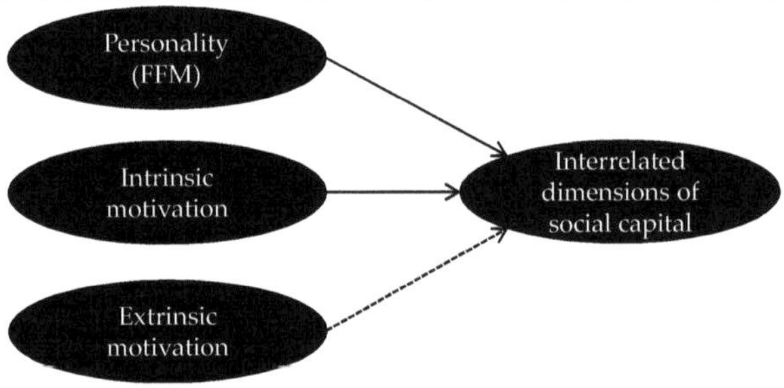

Solid lines represent significant relationships, the dashed line represents insignificant relationships.

Figure 5: Summarized findings of the SEM approach presented in part IV

Overall, the main contribution of the studies presented in this dissertation lies in the multifaceted examination of community network antecedents through the theoretical lens of social capital. As described above, the first study incorporates multiple dimensions as well as multiple levels of social capital into the analysis and offers propositions based on qualitative evidence. The second study builds on the first study and tests proposed relationships on the individual level on a quantitative basis. In this context, the two studies presented complement each other. The first study offers a holistic framework of antecedents of social capital in community settings, while the second study presents more finegrained analysis for scholars and practitioners who are interested in individual level antecedents of social capital. These aspects play a pivotal role in recruiting and incentivicing activities in community settings.

This dissertation is structured as follows. In the next part, I review relevant literature in the context of innovation communities and social capital. This second part ends with a presentation of the research gap as well as the research question this dissertation aims to answer. In the third and fourth part, I present the qualitative multiple case study and the quantitative covariance-based structural equations modeling (CBSEM) approach, respectively. I end with a general, collective discussion of both studies including theoretical and managerial implications, as well as limitations and avenues for future research.

II

Theoretical background:

Open innovation, communities, and social capital

1 Structure of part II

Part II reviews studies that are of particular importance in the context of this dissertation. As presented in part I, this dissertation takes the theoretical lens of social capital and looks at innovation communities as a means to organize boundary crossing innovation processes. In consequence, three literature streams are highly relevant: (1) The open innovation literature, which discusses boundary crossing innovation processes, (2) the innovation community literature, which looks at communities and their role with regard to innovation, and (3) the social capital literature, which provides the theoretical lens applied.

The second chapter of part II gives a general overview of the open innovation literature. This strand of research is important because many studies on innovation communities refer to open innovation scholars (e.g., Bullinger, Neyer, Rass, & Möslein, 2010; Füller et al., 2008). In essence, the open innovation literature provides insights into theoretical lines of arguments for advantages of boundary crossing innovation processes that are highly relevant in the context of innovation communities. Amongst other things, open innovation scholars discuss various groups of external stakeholders (e.g., suppliers, customers or universities) that are valuable partners for companies in innovation processes. The partners discussed in this stream of research often engage in communities in order to innovate (e.g., Jeppesen & Laursen, 2009).

The third chapter provides an overview of the innovation community literature[5]. The review is structured according to the definition of corporate innovation communities already presented in part I. As the chapter highlights, researchers have investigated innovation communities inside the boundaries of organizations (Bansemir et al., 2012), outside the boundaries of organizations (Schreier & Prügl, 2008) as well as mixed communities (Bjelland & Wood, 2008). Furthermore, the chapter provides an overview of studies that emphasize voluntary participation (e.g.,

[5] For a systematic literature review on innovation communities, see Bansemir (2012). While Bansemir (2012) focuses on communities within the boundaries of organizations, he also incorporates arguments of studies on innovation communities outside organizational boundaries.

Dahlander & Frederiksen, 2012), joint effort (e.g., Franke & Shah, 2003), and online (e.g., Bansemir et al., 2012) as well as offline (e.g., Bechky, 2003) interaction. Finally, supporting activities of companies in corporate innovation communities are discussed (e.g., di Gangi & Wasko, 2009).

The fourth chapter reviews the literature on social capital. It starts with the historical roots of the concept in the academic literature, followed by a definition as well as the relevance of social capital in the context of corporate innovation communities. Furthermore, refering to Nahapiet and Ghoshal (1998), the review provides an overview of extant research with regard to the structural, the relational, as well as the cognitive dimension of social capital. Thereafter, the multilevel nature of social capital is discussed (Payne et al., 2011). This part ends with the presentation of the research gap and the research question addressed in this dissertation.

2 The paradigm of open innovation

This chapter provides an overview of the open innovation literature, which examines innovation processes that integrate both internal as well as external partners to gain access to knowledge, ideas, and other resources (Chesbrough, 2003a; Enkel, Gassmann, & Chesbrough, 2009). As such, many scholars in the context of innovation communities refer to this research (Schröder & Hölzle, 2010; di Gangi & Wasko, 2009) and extant literature on the topic is highly relevant in the context of this dissertation. In more detail, a focal advantage of community-based innovation processes lies in the flexible exchange of knowledge and ideas across boundaries and open innovation scholars offer theoretical arguments for the advantages of such processes.

Of course, the idea that companies rely on knowledge developed outside their own R&D departments in order to innovate is not new. In essence, scholars have discussed the boundary crossing nature of innovation processes for decades (Trott & Hartmann, 2009). For example, technical innovation in blast furnace practice in the mid-nineteenth century in England was a result of intensive knowledge exchange between different companies (Allen, 1983). Likewise, Smith (1976) argues that many inventions during the age of industrialization have been put forward by users of machines. According to Smith (1976), the division of labor enabled individuals to focus on a specific part of a machine, and, thus, workmen were able to find out ways to improve their working equipment.

Similarly, the number of strategic alliances that aim to foster organizational innovation has increased at an enormous scale in the 1970's and 1980's (Hagedoorn & Schakenraad, 1994; Harrigan, 1988). Hagedoorn (1993) argues that motives and targets of strategic technology partnering are diverse and differ between industries. Companies in high-tech sectors form strategic alliances because of technology-related motives such as capturing technological knowledge and technology transfer. In contrast, companies in mature sectors co-operate in strategic technology partnerships mainly because of market-related objectives such as searching for new opportunities with respect to product line expansion or entry into international markets. Fang (2011) finds that strategic alliances that leverage complementary knowledge bases lead especially in those cases to innovative outputs when partners collaborate intensively as intense collaboration generates trust and familiarity.

Authors who study innovation processes that cross the boundaries of R&D departments and organizations in the recent decade often refer to the framework of open innovation (Almirall & Casadesus-Masanell, 2010). Chesbrough describes open innovation processes as a vision, which "eagerly seeks external knowledge and ideas, even as it nurtures internal ones" (2003a: xxxi). In open innovation processes, core inside innovators (i.e., R&D department employees) collaborate with peripheral inside innovators (i.e., employees from other departments), and external innovators (i.e., individuals who do not belong to the focal company) (Neyer, Bullinger, & Möslein, 2009). Those authors who refer to Chesbrough (2003a,b) argue that open innovation entails both inbound and outbound activities (Dahlander & Gann, 2010; Gassmann & Enkel, 2004). Inbound open innovation activities refer to the integration of external knowledge and ideas in innovation processes (Chesbrough & Crowther, 2006). In contrast, outbound open innovation activities refer to the external exploitation of internally developed knowledge, for example by means of technology licensing (e.g., Grindley & Teece, 1997). Furthermore, a number of scholars have referred to the notion of coupled open innovation to describe activities that entail both inbound as well as outbound processes simultaneously (Bogers, 2011; Enkel et al., 2009).

While some scholars express serious doubt that Chesbrough's paradigm of open innovation describes a completely new phenomenon (e.g., Trott & Hartmann, 2009), there is scientific consensus that boundary crossing innovation processes nowadays gain on greater significance. The opening of innovation processes is no longer a phenomenon which is limited to high-tech industries, but can be observed in low-tech industries such as home improvement or consumer packaged goods as well (Chesbrough & Crowther, 2006). In a similar vein, open approaches to innovation are not restricted to early stages of industry life cycles. Christensen, Olesen, and Kjar (2005) study open innovation processes in the consumer electronics industry and find that, in the embryonic stage of the industry, universities, high-tech start-ups and medium-sized high-end OEMs collaborate in order to innovate. In mature stages of the consumer electronics industry, to the contrary, large-scale semiconductor suppliers provide breakthrough complementary R&D inputs and play a crucial role in open innovation processes. Hence, companies organize open innovation approaches differently during industry evolution, but collaborate with external partners in each stage of the industry life cycle.

The open innovation literature discusses different groups of potential partners in open innovation processes. Open innovation scholars often emphasize the role of customers (Piller & Walcher, 2006), suppliers (Petersen, Handfield, & Ragatz, 2003), and universities (Bercovitz & Feldman, 2007) as collaboration partners in innovation processes[6]. These partners are at the same time valuable members of corporate innovation communities. Those scholars who examine the role of external partners in innovation processes often build on von Hippel's concept of sticky knowledge (von Hippel, 1994). The main argument of von Hippel is that external partners in many cases possess knowledge that is costly to transfer and, therefore, it is beneficial to integrate them directly in innovation processes (Lüthje, Herstatt, & von Hippel, 2005).

Scholars have examined innovative activities of users and customers in different industries, for example in the construction equipment industry (Herstatt & von Hippel, 1992) or in banking services (Oliveira & von Hippel, 2011). In this context, lead user theory has gained particular importance (von Hippel, 1986). Lead user theory suggests that users who expect the largest benefits from innovations are in many cases those users who experience needs earlier than the majority of their peers. Simultaneously, these users possess highly relevant knowledge for innovation processes (Lilien, Morrison, Searls, Sonnack, & von Hippel, 2002) and, hence, are promising partners in innovation activities. Foss, Laursen, and Pedersen (2010) argue that interaction with users is especially beneficial for innovative performance of companies when interaction with users is complemented by appropriate organizational practices such as delegation of decision rights and incentivizing knowledge sharing. As Foss et al. (2010) argue, these organizational practices enhance communication in the organization and ascertain that those who interact with customers share the obtained knowledge.

Because of their well-marked knowledge about single components of products, suppliers possess valuable knowledge in the context of boundary crossing innovation

[6] Open innovation partners discussed in this chapter are illustrative examples. The literature provides numerous other external innovation partners, for instance competitors (Gnyawali & Park, 2011).

activities, too (Takeishi, 2001). Lau, Tang, and Yam (2010), for instance, find that product co-development with suppliers is positively related to product innovation, which in turn fosters performance of products. Examining component development projects in the automobile industry, Takeishi (2001) finds evidence that intense problem solving with suppliers during early stages of a project leads to higher quality of the component design developed. In addition, Takeishi (2001) finds that face-to-face communication with suppliers is beneficial for component development projects.

Other researchers argue that collaboration with universities is associated with higher R&D performance (Asakawa, Nakamura, & Sawada, 2010). There is a multiplicity of benefits that potentially arise from interaction with universities in innovation processes, ranging from the absorption of latest scientific knowledge to concrete solving of technological problems (Salter & Martin, 2001). The relevance of the benefits of collaboration with universities differs between industries. For example, collaboration with university partners who possess latest scientific knowledge is most promising in the life science industry (Zucker, Darby, & Armstrong, 2002). To the contrary, in the advanced materials industry, collaboration with university researchers who are interested in both advanced scientific understanding as well as practical applicability leads to most successful innovation processes (Baba, Shichijo, & Sedita, 2009).

A distinguishing mark between earlier inter-firm innovation and latest open innovation approaches is that nowadays companies do not restrict innovative collaboration to a single external innovator (e.g., customer, supplier, or university), but integrate a diverse set of partners simultaneously in order to gain from synergistic effects (Bahemia & Squire, 2010). In their often cited paper, Laursen and Salter (2006) examine the relationship between the number of external partners (i.e., external search breadth) in innovation processes and innovative performance of companies and find a curvilinear relationship. In other words, initially companies benefit from higher numbers of external partners. However, companies may also over-search, and therefore, companies that cooperate with too many partners will suffer from lower innovative performance (Leiponen & Helfat, 2010). Related to that, companies nowadays often refrain from preselection of innovation partners, but rather publish an open call which describes the task to be accomplished and allows self-selection of qualified solvers of the problem at hand (Jeppesen & Lakhani, 2010).

Many studies highlight the role of information and communication technology (ICT) in recent trends in open innovation approaches (Dodgson, Gann, & Salter, 2006). While early approaches of user innovation relied on offline methods such as lead user workshops (e.g., Lilien et al., 2002), companies nowadays more and more take advantage of online settings (Baldwin & von Hippel, 2011). With the rise of ICT, innovation communities have received increased attention both in academia and practice because ICT allows organizations to integrate large cohorts of geographically dispersed individuals into innovation processes (di Gangi et al., 2010).

In a nutshell, open innovation scholars discuss innovation processes that cross the boundaries of departments and organizations. Within corporate innovation community processes, knowledge, ideas and other resources are exchanged across both departmental as well as organizational boundaries, too. Thus, innovation communities are one way to organize open innovation (Ili, Albers, & Miller, 2010). In consequence, literature strands on open innovation and innovation communities are highly interrelated and characterized by numerous cross citations (e.g., Bullinger et al., 2010; Franke & Shah, 2003; Füller et al., 2008). External partners mentioned in the open innovation literature are often discussed as potential and valuable members of innovation communities. For instance, many scientists are part of the innovation community *innocentive.com* (Jeppesen & Lakhani, 2010) and users and customers are among the most often discussed community members (Franke & Shah, 2003; Füller et al., 2008). Figure 6 summarizes my arguments. In the next chapter, I review studies on innovation communities.

What is open innovation?	The open innovation literature examines innovation processes that integrate both internal as well as external partners in innovation processes to gain access to knowledge, ideas, and other resources (Chesbrough, 2003a; Enkel et al., 2009; Neyer et al., 2009).		
Often discussed partners in open innovation	**Users and customers** Within this group, lead users who experience needs earlier than the majority have received particular attention because they possess both need as well as solution knowledge (von Hippel, 1986; Lilien et al., 2002).	**Suppliers** Suppliers are promising partners in innovation processes because they possess well-marked knowledge about single components of products (Takeishi, 2001; Lau et al., 2010).	**Universities** Potential contributions of universities in innovation processes range from exchange of latest scientific knowledge to concrete solving of technological problems (Salter & Martin, 2001; Baba et al., 2009).
Relevance for innovation communities	Innovation communities are one way to support open innovation processes. In consequence, literature strands on open innovation and innovation communities are highly interrelated (e.g., Bullinger et al., 2010; Franke & Shah, 2003; Füller et al., 2008).		

Figure 6: Open innovation and its relevance in innovation communities

3 The central characteristics of corporate innovation communities

Many scholars have examined the role of communities in knowledge sharing (Faraj, Jarvenpaa, & Majchrzak, 2011; Faraj & Johnson, 2011) and innovative activities (Bansemir et al., 2012; Franke & Shah, 2003; Franke, Keinz, & Schreier, 2008). A review of these studies exhibits that scholars differ considerably in their understanding of the term innovation community (Muhdi & Boutellier, 2011). The divergent conceptualizations of the community concept constrain scientists to clearly state their specific understanding of the term innovation community. As West and Lakhani put it: "We encourage researchers to explicitly articulate the theoretical and phenomenological boundaries of their use of the term, both to build upon earlier scholarship and to enable subsequent scholars to build on them" (2008: 224).

As stated in the introduction, I define corporate innovation communities as *heterogeneous* associations of actors who *voluntarily* perform innovative activities in a *joint effort* in both *online* (e.g., IT-based innovation platforms) as well as *offline* (e.g., workshops) environments. A *company supports* the activities of community members who may or *may* not *be employed* by the focal organization. I argue that this definition highlights five central characteristics of corporate innovation communities. Corporate innovation communities (1) cross the boundaries of units and organizations. Furthermore, members of corporate innovation communities engage (2) voluntarily and (3) in a joint effort in innovation processes. In addition, corporate innovation communities exchange knowledge and ideas both in (4) online as well as offline settings. Finally, (5) corporate innovation communities receive support by a company that is interested in the work of community members. Table 1 portraits propositions, a summary of extant literature as well as exemplary publications for each of the central characteristics.

Table 1: Central characteristics of corporate innovation communities

#	*Proposition*	*Summary of extant literature*	*Related Publications*
1	Corporate innovation communities cross the boundaries of units and organizations.	Extant research discusses innovative activities of communities within organizations (e.g., communities of practice) as well as of communities outside the boundaries of organizations (e.g., OSS communities).	Brown & Duguid (1991); Franke & Shah (2003); Lerner & Tirole (2002); Muhdi & Boutellier (2011)
2	Corporate innovation community members act voluntarily.	Because of the voluntary nature of innovation communities, many scholars have examined motivations of community members to engage in innovative activities. The literature distinguishes between intrinsic and extrinsic motivations.	Franke & Shah (2003); Lerner & Tirole (2002); Nambisan & Baron (2010); von Hippel & von Krogh (2003)
3	Corporate innovation community members act in a joint effort.	Researchers stress the intertwined and collaborative efforts of community members when working together.	Bechky (2003); Franke & Shah (2003); Hislop (2003); Muhdi & Boutellier (2011)
4	Corporate innovation community members exchange knowledge in offline as well as online settings.	Scholars report both innovative activities of communities in offline as well as online settings.	Bansemir et al. (2012); Brown & Duguid (1991); McLure Wasko & Faraj (2000); Stam (2009)
5	Corporate innovation communities receive corporate support.	Many scholars have described how companies support innovative activities of communities.	Dahlander & Frederiksen (2012); di Gangi et al. (2010)

First, I argue that innovation communities cross the boundaries of units and organizations. In doing so, I refer to Muhdi and Boutellier (2011) and argue that innovation communities may be located inside as well as outside the boundaries of organizations[7]. My arguments are closely related to those of Neyer and colleagues (2009), who separate between three groups of people in open innovation processes: (a) core inside innovators (i.e., the R&D department), (b) peripheral innovators (i.e.,

[7] Note that characteristic #1 in Table 1 is different from characteristic #5. That is, internal as well as external communities may receive corporate support.

employees who do not innovate according to their job descriptions, but are interested in and able to produce innovative ideas), and (c) outside innovators (i.e., all external innovators who are not members of the focal company). I argue that a group of people solely consisting of members of a single department is not an innovation community as innovation communities operate outside hierarchical structures (Fleming & Waguespack, 2007) and are of heterogeneous nature. An analysis of extant literature on innovation communities reveals that some scholars examine innovative activities of communities inside the boundaries of organizations (e.g., Bansemir et al., 2012), while others focus on external communities (e.g., West & Lakhani, 2008).

An early research stream which explored primarily internal innovation communities looks at communities of practice (Wenger & Snyder, 2000). Innovation community scholars often refer to studies on communities of practice as the underlying arguments of both research strands are similar. In their seminal work on communities of practice, Brown and Duguid (1991) propose that the processes of working, learning, and innovating in organizations are better portrayed by the concept of communities of practice as opposed to job descriptions and organizational charts. In more detail, they argue that organizational processes are highly complex and characterized by intertwined activities between individuals. Hence, according to Brown and Duguid (1991), these processes cannot be described in simple canonical steps. Furthermore, learning and innovating are not conflicting activities to working. In contrast, people learn and innovate whilst working together (Hislop, 2003). Thus, working, learning, and innovating are social processes that describe the advances of groups of people (i.e., communities of practice) in understanding how they best work and innovate together.

More recently, scholars have examined the role of communities situated outside the boundaries of organizations in innovation processes (Füller et al., 2008; Nambisan, 2002), often with a specific focus on sports (Hienerth, 2006; Piller & Walcher, 2006; Schreier & Prügl, 2008) or OSS communities (Lerner & Tirole, 2002). Franke and Shah (2003), for instance, investigate innovative activities of members of canyoning, boardercross, sailplaning, and handicapped cycling communities and find that almost a third of all community members had already innovated. Among others, Lakhani and von Hippel (2003) have examined innovation processes within OSS communities. My definition of corporate innovation communites allows participation of internal and external members simultaneousy.

Second, I stress the voluntary nature of community engagement. Innovation community members act in environments without clearly defined job descriptions and formal structures (Brown & Duguid, 1991; Fleming & Waguespack, 2007). As a result, hierarchical authority plays a minimized role in community settings (Judge et al., 1997; Muhdi & Boutellier, 2011). Community members engage in innovative activities because of their own will (Nambisan, 2002). Related to that, researchers have examined motivations of community members to engage voluntarily in innovative activities (e.g., von Hippel & von Krogh, 2003). A popular categorization of motivational aspects is provided by Deci (1972) who distinguishes between intrinsic motivation directly related to the focal activity and extrinsic motivation related to some form of external rewards. A number of researchers have used the distinction between intrinsic and extrinsic motivations to structure motivations of innovation community members (Janzik, 2010)[8]. With respect to extrinsic motives, many scholars have found evidence that learning from others (Lakhani & von Hippel, 2003) as well as expertise enhancement (Nambisan & Baron, 2010) reinforce contribution behavior in innovation communities. Examples for intrinsic motivations are, for instance, a feeling of responsibility to the community (Nambisan & Baron, 2010) or enjoyment in the joint work with other community members (Franke & Shah, 2003). Lerner and Tirole (2002) argue that individuals might be motivated to contribute to communities because tasks accomplished within communities result in more fun than their daily routine tasks. Similarly, McLure Wasko and Faraj (2000) propose that moral obligation toward the focal community constitutes a major motivational factor in the context of contribution behavior within innovation communities.

Third, a common theme of corporate innovation communities is the collaborative and joint effort of members in knowledge exchange and innovating activities (Franke & Shah, 2003). Researchers of communities of practice stress the intertwined and collaborative efforts of employees when working together (Bechky,

[8] See part IV of this dissertation for a more detailed discussion of extrinsic and intrinsic motivations within innovation communities.

2003). Among others, Hislop stresses the "group based character of organizational activity" (2003: 165), thereby pointing to intense knowledge exchange between individuals. In a similar vein, Franke and Shah find that "user innovation is not an individual task but a joint effort" (2003: 164). Some of the community-based motivations to engage in innovative activities discussed above are related to the interactive and collaborative way of working in innovation communities. In their recent study, Muhdi and Boutellier (2011) find that the development of ideas together with other community members constitutes an important motivational factor in an internal innovation community of a Swiss bank. Hertel, Niedner, and Herrmann (2003) find that members of OSS communities rarely develop code on their own, but often cooperate with other members in spontaneous teams. In essence, the arguments of scientists who stress the intertwined nature of innovative activities relate to those of Nahapiet and Ghoshal (1998), who argued more than a decade ago that there is consensus among scholars that innovations come into being as individuals combine different stocks of knowledge and ideas.

Fourth, I stress that innovation communities exchange knowledge and ideas in offline settings (Brown & Duguid, 1991; Füller et al., 2008), as well as on online platforms (Bansemir et al., 2012; Faraj & Johnson, 2011; Nambisan, 2002). Bechky (2003), for instances, discusses knowledge exchange processes between three communities of practice (i.e., engineers, assemblers, and technicians) in an offline setting. She argues that knowledge exchange between different occupational communities is difficult as different communities use different languages. In addition, Bechky (2003) argues that technicians are able to connect the two other occupational communities because technicians are conversant with both languages of engineers as well as assemblers. With respect to online interaction, McLure Wasko and Faraj (2000) examine electronic communities of practice and propose that knowledge exchange within these communities works best if knowledge is viewed as embedded within the community. Stam (2009) argues that online and offline knowledge exchange complement each other and finds that companies benefit especially from online participation in an OSS community when these companies engage in offline activities (e.g., organization of workshops) at the same time. User communities of the Danish toy manufacturer LEGO are an example for simultaneous knowledge and idea exchange on online platforms as well as offline events (Frey & Lüthje, 2011).

Fifth, corporate innovation communities are supported by organizations that are interested in the outcome of community work. This understanding of an innovation community is closely related to the arguments of Agerfalk and Fitzgerald (2008), who examine how software companies outsource software development to global OSS communities. In essence, a number of scholars have discussed corporate ways to benefit from the creative potential of innovation communities (Judge et al., 1997; von Hippel, 2001) and some have suggested that an opportunity is to support the establishment of a corporate community (Sawhney & Prandelli, 2000). Examples for companies that support innovation communities are numerous and exist in diverse industries, such as IT (Pisano & Verganti, 2008), musical instruments (Dahlander & Frederiksen, 2012), or toys (Frey & Lüthje, 2011). Within the communities studied in this dissertation, employees from different departments of the supporting organization innovate together. Furthermore, some communities I examine also allow individuals who do not belong to the focal organization to join the community[9]. These facets of the innovation communities investigated in my dissertation are similar to the IBM innovation jam initiative (Bjelland & Wood, 2008).

So far, I have reviewed extant literature on open innovation activities in general and on innovation communities in particular. Many of the studies cited above stress the importance of social capital in open innovation and innovation communities (Fey & Birkinshaw, 2005; Fleming & Waguespack, 2007). Boundary crossing innovation processes comprise knowledge and idea exchange between individuals situated in different departments and organizations (Neyer et al., 2009). In such settings, social relationships affect knowledge and idea exchange as well as innovative performance significantly (Ahuja, 2000; Tsai, 2000; Tsai & Ghoshal, 1998). I review extant studies on social capital with a specific focus on innovative activities in the next chapter.

[9] The characteristics of each innovation community examined are described in detail in the methodological sections.

4 The relevance of social capital in innovation processes

In this chapter, I review the literature on social capital. The core argument of social capital traces back to social network theory and Granovetter, who suggested almost two decades ago that all "attempts [of individuals or organizations] at purposive action are [...] embedded in concrete, ongoing systems of social relations" (1985: 487). Influenced by this perception, scholars on social networks have analysed which network structures and positions are most promising. The discussion around this question in the literature is extensive and sketched through two opposing views. The first view has been put forward by Coleman (1988) who argues that cohesive networks are beneficial as cohesion is associated with trusted exchange between network partners. Gargiulo and Benassi summarize Coleman's (1988, 1990) arguments: "[M]embers of a closely knit network can trust each other to honor obligations, which diminishes the uncertainty of their exchanges and enhances their ability to cooperate in the pursuit of their interests" (2000: 184). The second and opposing view has been put forward by Burt (1992) who argues that structural holes in networks are advantageous as sparse network structures lead to brokerage opportunities. In other words, a network member who is in contact with otherwise disconnected members benefits from filtering and controlling information and other resources (Burt, 1997). The intense discussion of these two groups, characterized by contradicting theoretical arguments and empirical findings, has brought a third group of scholars to the scene. This third group looks at environmental moderators that reinforce or mitigate the benefits and drawbacks of both cohesive networks as well as structural holes (e.g., Maurer & Ebers, 2006; Rodan & Galunic, 2004).

The concept of social capital, as it is used today in the management literature, is closely connected to the research of Nahapiet and Ghoshal (1998) who argue that analyzing social capital urges for more than the mere analyses of network structures. In the following, I present popular definitions of social capital and highlight its importance in different settings with a focus on innovation processes.

A review of the literature on networks and social capital shows that the concept is complex (Granovetter, 1973) and entails a high number of heterogeneous facets. The research presented in this dissertation aims to consider the complex nature of social capital in innovation communities. Researchers have repeatedly highlighted that in order to draw a holistic picture of social capital, scholars need to regard both multiple dimensions (Bolino et al., 2002; Hughes & Perrons, 2011) as well as multiple levels (Leana & Pil, 2006; Oh, Labianca, & Chung, 2006) in their analysis. Responding to their calls, I discuss these two facets of social capital in the second and third section of this chapter in order to provide a theoretical basis for the understanding of social capital in this dissertation. First, as Nahapiet and Ghoshal (1998) have raised, social capital research demands analysis of a structural, a relational, and a cognitive dimension of social relationships and networks. In essence, the inclusion of the second and third dimension is a distinguishing mark between social network theory and the concept of social capital in the management literature. Second, I follow Payne and others (2011) who stress the multilevel characteristic of social capital.

4.1 Social capital: Definition and relevance

In line with the tradition of social network research, social capital scientists also emphasize the importance of social ties and propose that individuals and collectives benefit from their network of relationships as they gain access to resources through these networks (Blyler & Coff, 2003). Moreover, Inkpen and Tsang (2005) conclude that a network itself preoccupies the characteristics of a resource. In their often cited review on social capital, Adler and Kwon define social capital as "the goodwill available to individuals or groups. Its source lies in the structure and content of the actor's social relations. Its effects flow from the information, influence, and solidarity it makes available to the actor" (2002: 23). Similarly, Nahapiet and Ghoshal state that social capital is "the sum of the actual and potential resources embedded within, available through, and derived from the network of relationships possessed by an individual or social unit" (1998: 243).

Extant scholars have demonstrated the importance of social capital in different research settings. In this context, social capital is often framed as a central resource that affects how organizations advance to reach strategic goals (Lovas & Ghoshal, 2000). For instance, a number of scholars have investigated whether and how social capital affects performance of organizations. Some, for example Uzzi (1999) and Batjargal and Liu (2004), find that social capital influences access of organizations to financial resources. Others have analyzed the role of social capital in the context of organizational dissolution (Polidoro Jr., Ahuja, & Mitchell, 2011). As an example, Pennings, Lee, and van Witteloostuijn (1998) find that social capital that connects a company to its clients is negatively related to firm dissolution.

The importance of social capital in the context of economic activities is also reflected in studies which rely on social capital theory in order to examine consequences of employee mobility or turnover (Dess & Shaw, 2001; Shaw, Duffy, Johnson, & Lockhart, 2005). In this context, Somaya, Williamson, and Lorinkova (2008) find that employee transfer between a company and its clients is beneficial for firm performance (irrelevant of the direction of exchange). In contrast, employee transfer from competitors is beneficial for firm performance, while employee transfer to competitiors is harmful for firm performance. Related to that, Dokko and Rosenkopf find that companies can increase their influence in standard setting committees by hiring people that possess valuable social capital and conclude that "[a]s well as, or instead of, building social capital through the process of interaction over time, firms can use hiring as a shortcut to increase their social capital" (2010: 689).

Researchers on multinational enterprises and international business activities have also shown that social capital affects business activities significantly (Kostova & Roth, 2003). Li, Poppo, and Zhou (2008) emphasize the importance of social capital and managerial ties for businesses in China. They find that while managerial ties are beneficial for domestic businesses, foreign businesses experience decreasing returns from managerial ties. Similarly, Shaner and Maznevski (2011) investigate the consequences of social relationships in the context of foreign direct investments and find that ties to direct profit generating contacts, such as customers or suppliers, are especially valuable for country business units of multinational enterprises.

In addition, scholars have demonstrated the implications of social capital for non-profit organizations. Galaskiewicz, Bielefeld, and Dowell (2006), for instance, find that donative non-profit organizations (i.e., non-profits that rely on donors and volunteers) grow faster if these organizations own a central position in interorganizational networks. As another example, Pil and Leana (2009) find that social capital of school teachers affects student performance improvement.

The previous three paragraphs highlight the importance of social capital for economic activities in general. Beside the research fields mentioned above, scholars have used a social capital lens to investigate three interrelated contexts which are of particular importance for this dissertation. First, scholars have argued that social capital is beneficial for knowledge exchange (Cross & Sproull, 2004; Reagans & McEvily, 2003) and knowledge creation (Lavie & Drori, 2012; McFadyen & Cannella Jr., 2004). An often cited study in this context has been published by Hansen (1999), who shows how closure and frequency of interaction via social ties (i.e., tie strength) interacts with successful knowledge exchange. Hansen (1999) finds that while weak ties are advantageous in the search for useful knowledge, strong ties are outstanding when it comes to transfer of that knowledge. In a related study, McEvily, Jaffee, and Tortoriello find that "brokerage benefits derived from bridging ties are not limited to timely access to information in the short-term but can also include the persistent effect of learning and knowledge redeployment" (2012: 59). Furthermore, McEvily et al. (2012) find that especially those bridging ties that are established during early career stages are important because they are associated with learning. Proponents of the central role of social capital in knowledge exchange processes often argue that knowledge exchange takes place by chance and voluntary and that hierarchical pressure (Tsai, 2002) and authority (Inkpen & Tsang, 2005) are harmful in this context. As Inkpen and Tsang put it: "[H]eadquarters of an intracorporate network must decentralize authority to members of the network so that they can determine how to make the best use of the knowledge they possess" (2005: 156).

Second, researchers have proposed that social capital affects creativity (Lingo & O'Mahony, 2010; Perry-Smith & Shalley, 2003). Among others, Perry-Smith (2006) argues for the importance of social ties within creative processes and finds that the relationship between weak ties and creativity of individuals is partly mediated by background heterogeneity of interacting network partners. Sosa (2011) examines the role of social capital in the context of dyadic interactions and finds that breadth of knowledge domains acquired as well as tie strength between interacting individuals is positively related to the generation of creative ideas[10]. With regard to network cohesion, Sosa (2011) identifies a concave relationship with creative idea generation. In other words, while network density does not affect idea generation in the context of low levels of cohesion, it significantly and negatively affects idea generation at high levels of cohesion.

Finally, extant literature often highlights the role of social capital in innovation processes (Bouty, 2000; Hansen, Mors, & Løvås, 2005; Laursen, Masciarelli, & Prencipe, 2011; Phelps, 2010; Rowley, Behrens, & Krackhardt, 2000). This argument is closely related to the research findings reviewed in the two previous paragraphs. Innovation processes are often described as processes of knowledge exchange and combination (e.g., Bercovitz & Feldman, 2007). Creativity plays an important role in innovation processes, especially during early stages (i.e., the fuzzy front end of innovation) (e.g., van de Ven, 1986). Subramaniam and Youndt summarize the relevance of social capital in innovation processes as follows: "[U]nless individual knowledge is networked, shared, and channeled through relationships, it provides little benefit to organizations in terms of innovative capabilities" (2005: 459). In their study, they find that while human capital (i.e., skills, capabilities, or experiences) alone is negatively related to radical innovation, the interaction of human and social capital is positively related to radical innovation. In addition, Subramaniam and Youndt (2005) find that social capital is positively related to incremental as well as radical innovation. Relatedly, Maurer, Bartsch, and Ebers (2011) find that trusted, strong ties in intra-organizational networks foster knowledge transfer and innovative performance of companies. Lovejoy and Sinha propose that, in the context of innovative activities, "ideation is accelerated by a "complete graph" type network, which maximizes the number of parallel conversations per period while encouraging people to dynamically churn through a large (ideally the entire population) set of conversational partners" (2010: 1139). Tsai and Ghoshal (1998) find that the relationship between the three dimensions of social capital and product innovation is mediated by resource exchange and combination. Furthermore, researchers have

---

[10] Obviously, the findings of Perry-Smith (2006) and Sosa (2011) are conflicting. I discuss different theoretical lines of argument as well as empirical findings with regard to the consequences of social capital in the next section in detail. Here, I solely argue for the relevance of social capital in innovation processes.

shown that social capital significantly affects the likelihood of successful innovation processes in the context of regional clusters (Whittington, Owen-Smith, & Powell, 2009) as well as strategic alliances (Ireland, Hitt, & Vaidyanath, 2002).

Briefly, social capital provides researchers a lens to investigate how social relationships allow resource exchange between network partners and are therefore useful for a single network partner or for the network as a whole. While a number of scholars have shown that social capital affects various measures of organizational performance, innovation processes are one central context in which researchers have used the concept. Figure 7 summarizes my arguments.

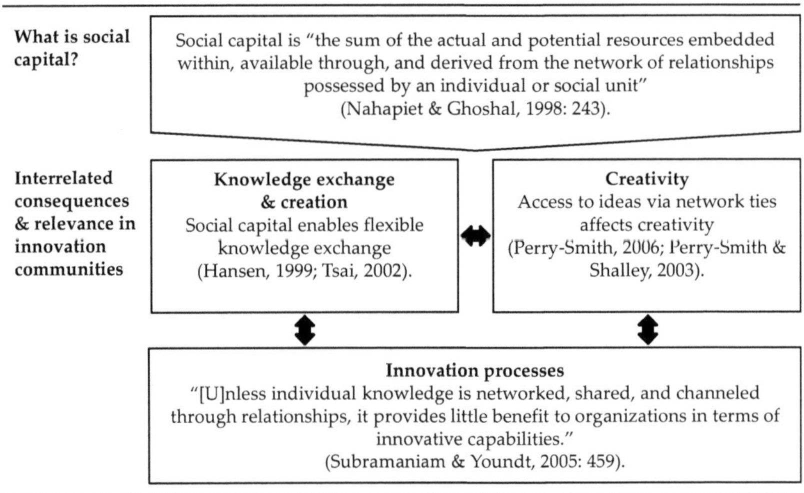

Figure 7: Definition of social capital and its relevance in innovation communities

As I have already mentioned, extant social capital scholars have repeatedly emphasized the complex nature of social capital and have called for research efforts that take care of the complexity of the concept (Hughes & Perrons, 2011; Oh et al., 2006). In order to follow the recommendations in the literature, I review two facets often mentioned with regard to the complexity of social capital: (1) Its multidimensional and (2) its multilevel characteristic. First, as Nahapiet and Ghoshal (1998) argue with regard to multiple dimensions, social capital is composed of a structural, a relational, and a cognitive dimension. Second, as emphasized by Payne et al. (2011), social capital is a multilevel concept.

4.2 The three dimensions of social capital

As reflected in the definition provided by Adler and Kwon (2002), researchers have repeatedly emphasized the multidimensional nature of social capital (e.g., Koka & Prescott, 2002). In their seminal article, Nahapiet and Ghoshal (1998) distinguish between three dimensions of social capital, a conceptualization of the concept that many scholars have referred to (e.g., Inkpen & Tsang, 2005; Tsai & Ghoshal, 1998). First, the structural dimension refers to the pattern of relationships between individuals or collectives in the network (e.g., Ahuja, 2000) and determines who is connected to whom (e.g., Tsai, 2001). The second dimension, coined relational dimension, describes the quality of relationships in the network (Lechner, Frankenberger, & Floyd, 2010). For example, the relational dimension captures the degree of trust and tie strength (Maurer et al., 2011). Finally, the cognitive dimension "refers to those resources providing shared representations, interpretations, and systems of meaning among parties" (Nahapiet & Ghoshal, 1998: 244) and entails for example a shared vision (e.g., Lechner et al., 2010) or language (e.g., Bolino et al., 2002) within the focal network. I summarize the research findings for each dimension of social capital in the following, starting with the most prevalent structural dimension, followed by the relational dimension as well as the cognitive dimension. In doing so, I follow scholars who argue for the consideration of multiple dimensions in the analysis of social capital (e.g., Bolino et al., 2002) and provide a theoretical basis for subsequent empirical analysis. Figure 8 gives an overview of the arguments presented.

Structural dimension

- Refers to the pattern of relationships between individuals in the network (e.g., Ahuja, 2000) and describes who is connected to whom (e.g., Tsai, 2001)
- While many ties (i.e., network size or centrality) are mostly described as beneficial for a single network member (e.g., Phelps et al., 2012; Tsai, 2001), analysis of the benefits and drawbacks of cohesive structures of whole networks has produced conflicting results (e.g., Burt, 2007; Obstfeld, 2005)

Relational dimension

- Describes the quality of relationships in the network (e.g., Lechner et al., 2010) and captures aspects of trust and tie strength (e.g., Maurer et al., 2011)
- Strong and trusted ties are associated with benefits as they facilitate knowledge and resource exchange (e.g., Chua et al., 2008; Moldoveanu & Baum, 2011) as well as with drawbacks as they are associated with redundant resources (e.g., Ferriani et al., 2009; Hansen, 1999)

Cognitive dimension

- "[R]efers to those resources providing shared representations, interpretations, and systems of meaning among parties" (Nahapiet & Ghoshal, 1998: 244) and captures aspects of a shared vision (e.g., Lechner et al., 2010) and language (e.g., Bolino et al., 2002)
- Cognitive social capital affects knowledge sharing positively (Collins & Smith, 2006), especially in uncertain and turbulent environments (Hughes & Perrons, 2011; Kirsch et al., 2010)

Figure 8: The three dimensions of social capital

4.2.1 The structural dimension of social capital

As suggested in the previous section, "the structural aspect of social capital involves examination of the extent to which [...] [network members] are connected" (Bolino et al., 2002: 510). As such, measures used for the structural dimension often refer to the number of ties existing in a network and researchers have focused on different aspects depending on their research question. More specifically, scholars who have analysed benefits of single network members have looked at network size or centrality, while scholars who have analysed structures of whole networks have used the measure of network cohesion or density (see Carpenter, Li, and Jiang (2012) for more details). I review extant research with regard to these two measures in the following.

The first aspect, network centrality, is regularly measured as the number of network ties a network member has (e.g., Lechner et al., 2010). Other scholars have used the term network size for the same measure (e.g., Wong & Boh, 2010). Most authors have described big networks and centrality as promising (e.g., Phelps, Heidl, & Wadhwa, 2012). For instance, Tsai (2001) finds that network centrality of a business unit in an intraorganizational network is positively related to innovation performance and that this relationship is moderated by the focal units absorptive capacity. As another example, Mahmood, Zhu, and Zajac (2011) find that centrality in buyer-supplier networks is positively associated with the development of innovative capabilities.

Those researchers who look at structures of whole networks and examine network cohesiveness or density have produced contradicting findings. In general, extant scholars have followed three lines of argument: (1) A first group of studies follows the arguments of Coleman (1988) and proposes that dense networks are beneficial; (2) a second group, to the contrary, follows the arguments of Burt (1997) and suggests that sparse networks are beneficial; and (3) a third group examines what environmental effects moderate the relationship between network cohesion and performance. I review illustrative studies for each of the three groups in the following.

An often cited proponent of the first view is Obstfeld (2005) who introduces the term tertius iungens orientation to refer to a tendency of individuals to create and facilitate ties and interaction between disconnected people in one's network. Obstfeld (2005) finds that the tertius iungens orientation as well as dense networks are positively related to involvement in innovation. Other examples are Reagans and McEvily (2003) who find that dense networks are positively related to ease of knowledge transfer and Tortoriello, Reagans, and McEvily (2012) who find that network cohesion facilitates knowledge acquisition in cross-unit knowledge tranfer. Similarly, Tortoriello and Krackhardt refer to the concept of Simmelian ties and state: "[A] tie becomes Simmelian when the parties involved in it are reciprocally connected to one other and each is reciprocally connected to another, third party" (2010: 170). As such, the concept of Simmelian ties is closely related to network density. Tortoriello and Krackhardt (2010) find a positive relationship between bridging intraorganizational boundaries with Simmelian ties and the generation of innovation.

Another prominent example is Ahuja (2000) who finds that structural holes have a negative influence on innovative performance. Similarly, Baba and Walsh (2010) find that dense networks with strong ties are beneficial for the successful introduction of breakthrough innovations in companies.

With regard to the advantages of sparse networks (second group studies), Burt (1997, 2007) himself as well as other scholars (e.g., McFadyen, Semadeni, & Cannella Jr., 2009) have published a number of papers. For example, Burt (2004) finds evidence for the benefits of structural holes as he shows that individuals whose networks span structural holes are more successful, for example with respect to quality of ideas. In another article, Burt (2007) introduces the notion of secondhand brokerage to provide a more nuanced view on the performance implications of structural holes. In more detail, Burt (2007) shows that the mere analysis of the direct contacts of an individual is not sufficient to predict performance of an individual. In contrast, the ways these direct contacts are related to each other via other indirect contacts are important with respect to the benefits of structural holes as these relationships determine information benefits of a focal actor. Other scholars who have argued for the superiority of sparse networks are, for example, Gargiulo and Benassi (2000) who find that networks characterized by structural holes enable managers to adapt to changes in the environment.

The third group of studies mentioned earlier arose from the contradicting theoretical arguments of the two opposing groups of studies reviewed in the previous paragraphs. Scholars who belong to this third group propose that network cohesion is not per se good or bad, but that the benefits and drawbacks of both structures depend on environmental moderators (e.g., Hite & Hesterly, 2001). For instance, Xiao and Tsui's (2007) findings suggest that the benefits of structural holes are contingent upon cultural context and are not applicable to cultures that are collectivistic[11]. Mors (2010) finds that partners of consulting firms who operate in homogeneous contexts benefit most from sparse networks as these network structures are associated with non-redundant information and, as a consequence, with innovation performance. In contrast, dense networks are beneficial for partners whose networks span across

[11] Note that others have shown that the benefits of structural holes are significant across different countries (Burt, Hogarth, & Michaud, 2000).

geographic and firm boundaries. In such circumstances, dense networks provide a suitable context to integrate the knowledge acquired. As another example, Gargiulo, Ertug, and Galunic (2009) propose that the implications of network density depend on the role of an individual in a certain network. For instance, while network density is beneficial for an individual who acts as an acquirer of information, it is harmful for an individual who acts as a provider of information. Maurer and Ebers (2006) suggest that while cohesive networks are beneficial in start-up phases for entrepreneurial firms, the same network structures become a liability in subsequent business development phases. In result, Maurer and Ebers (2006) argue for the importance of a relationship management that allows entrepreneurs to overcome social capital lock-ins. Rodan and Galunic (2004) find that while sparse networks and knowledge heterogeneity affect overall performance of managers equally and complementary, sparse networks are only beneficial with regard to innovative performance when they are accompanied by heterogeneous knowledge.

4.2.2 The relational dimension of social capital

The relational dimension of social capital captures the quality or the content of relationships as opposed to the structure of networks and plays a pivotal role in the context of innovation processes, too (Moran, 2005). Within this dimension, scholars have addressed different topics, for instance trust and tie strength (Maurer et al., 2011).

Scholars who have emphasized the importance of trust often refer to Nahapiet and Ghoshal (1998) and propose that trusted ties make knowledge sharing and information flow more easy (Moldoveanu & Baum, 2011) and, hence, are beneficial for innovation (Molina-Morales & Martínez-Fernández, 2009). Chua, Ingram, and Morris (2008) distinguish between cognition-based trust (i.e., trust from the head) and affect-based trust (i.e., trust from the heart) and find that while cognition-based trusted ties are associated with task advice and exchange of economic resources, affect-based trusted ties are associated with friendships. In addition, both types of trust are positively associated with ties who give career advices. Another example is the study of de Jong and Elfring (2010) who find that the relationship between intrateam trust and performance is partially mediated by team monitoring and team effort. Ferriani, Cattani, and Baden-Fuller (2009) examine the role of social capital in project entrepreneurship in the film industry. With regard to the relational dimension, Ferriani et al. (2009) find that a team composed of both known partners as well as newcomers is advantageous because moderate levels of trust within a project team are benefical for success. Bruneel, D'Este, and Salter (2010) explore the relational dimension of social capital in open innovation contexts and find that trust between partners is beneficial as it reduces collaboration barriers. Similarly, Bouty (2000) describes the importance of relational social capital (i.e., trust and acquaintance) in the decision of R&D managers to exchange resources with peers employed by other organizations.

The relational dimension also captures the strength of ties (Maurer et al., 2011; Moran, 2005), which is often described as closeness and frequency of interaction (Hansen, 1999; Lechner et al., 2010). As such, tie strength is also interrelated with trust (Levin & Cross, 2004). While the seminal work of Granovetter (1973) emphasizes the benefits of weak ties (i.e., access to non-redundant knowledge and other resources), other scholars have presented arguments for positive aspects of strong ties as well (Hansen, 1999). For instance, interorganizational strong ties are beneficial as these relationships allow organizations to adapt successfully to environmental changes (Kraatz, 1998). Levin and Cross (2004) find that tie strength is associated with access to useful knowledge via two distinct theoretical mechanisms. First, tie strength is negatively related to the receipt of useful knowledge as weak ties are associated with non-redundant information. Second, strong ties are positively associated with trust and trust is associated with receipt of useful knowledge.

4.2.3 The cognitive dimension of social capital

As I have already described above, the cognitive dimension of social capital captures shared visions (Nahapiet & Ghoshal, 1998) and involves a shared language (McLure Wasko & Faraj, 2005) within a network. While the influence of cognitive social capital on economic performance has been supported in extant studies (e.g., Lechner et al., 2010), scholars have continued to neglect this third facet of social capital to a large extent (Hughes & Perrons, 2011). Recently, Hughes and Perrons (2011) stressed the importance of the cognitive dimension, especially in situations when networks have to go through turbulent phases. They argue that "firm[s] can [...] protect its social capital under these circumstances by clearly defining norms and shared expectations with other firms in the network, and by maintaining a high degree of transparency and openness with stakeholders" (Hughes & Perrons, 2011: 170). Cognitive social capital also plays a significant role in the context of knowledge sharing (Collins & Smith, 2006). For instance, Haeussler (2011) finds that a shared vision and norm of open science, which conceptualizes information as a common good, is beneficial for knowledge sharing among industrial as well as academic researchers. Kirsch, Ko, and Haney (2010) stress the importance of cognitive social capital in settings where the output of a certain activity cannot easily be managed and where knowledge about the working processes is imperfect. As Kirsch and his coauthors (2010) argue, shared visions of the people involved in working processes in such settings are prized because cognitive social capital substitutes formal control mechanisms such as performance targets to a certain extent. Aspects of cognitive social capital have also been highlighted as motivational aspects in the context of communities of practice. For instance, McLure Wasko and Faraj state: "[P]eople in these communities are behaving prosocially and out of a sense of moral duty. People in these communities feel that sharing knowledge and helping others is 'the right thing to do', and people also have a desire to advance the community as a whole" (2000: 169).

4.3 Social capital – a multilevel concept

In addition to the importance of three dimensions, social capital is a multilevel concept (Payne et al., 2011). That is, the concept can be used to analyze relationships between individuals, groups, and organizations (Brass et al., 2004). The multilevel nature of social capital is also reflected in the following citation of Moliterno and Mahony: "Organizations are multilevel systems of relationships [...], and as a result, a network theory of organizations should similarly be multilevel in its scope and perspective" (2011: 444). In their argument, Moliterno and Mahony (2011) use the concept of systems of nested networks (Harary & Batell, 1981) and highlight that each company in a network of organizations is itself a network in lower levels of analysis, for instance on group or individual levels. Furthermore, both constructs as well as relationships between constructs need to be isomorphic across different levels of analysis.

While there is no doubt that individual level social capital can result in higher levels of social capital (Berends, van Burg, & van Raaij, 2011), it is fundamental to note that social capital between groups or organizations is more than the representation of the relationships between individuals that belong to the focal group or organization. This aspect is discussed by Barden and Mitchell (2007) who compare the influences of exchange experiences on both organizational as well as individual levels on subsequent exchanges. Relying on the concept of relational embeddedness, Barden and Mitchell find that "interorganizational exchange ties have influences on interorganizational exchange behavior that go above and beyond the influences of leaders' exchange ties" (2007: 1455).

The multilevel characteristic of social capital is of central importance for my dissertation. Referring to the notion of systems of nested networks (Harary & Batell, 1981; Moliterno & Mahony, 2011), I argue that a study on antecedents of corporate innovation community social capital should consider all three levels of social capital discussed in the literature. First, as an innovation community is composed of individuals, the individual level is obvioulsy of relevance. Second, as people act in a joint effort (Franke & Shah, 2003) and often form groups (Bullinger et al., 2010) while innovating within communities, the group level is also germane. Finally, because corporate innovation communities receive organizational support and involve members of the supporting organization (Agerfalk & Fitzgerald, 2008; Bjelland & Wood, 2008), the organizational level of social capital should also be included in a study on antecedents of corporate innovation community social capital. Figure 9 highlights my arguments.

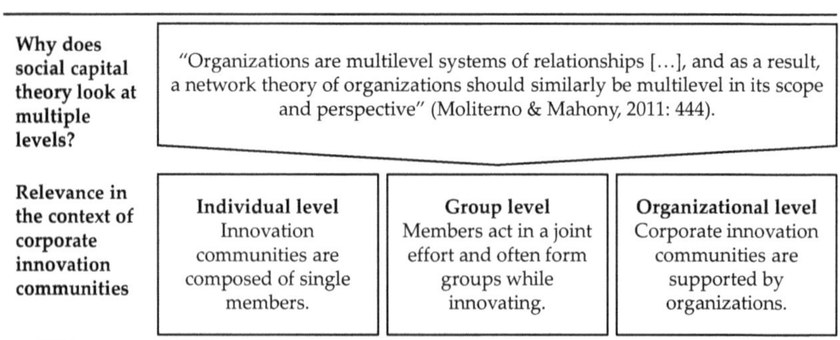

Figure 9: The relevance of the multilevel nature of social capital

In the following, I review illustrative social capital studies on individual, group, and organizational levels. Furthermore, I present exemplary studies that include cross-level analysis of social capital. The following review is important as it offers a theoretical understanding of the multilevel characteristic of social capital. As discussed in the next chapter 5, where I derivate my research question, this facet is central in the context of this research because scholars have not adressed this aspect enough in the past.

4.3.1 Individual level studies

A number of scholars have examined the relationship between individual level social capital and different performance measures. For instance, Belliveau (2005) finds that social capital affects the success of job search of women. In more detail, women who consult male ties were offered more jobs. A related study of Bian (1997) finds that chinese job seekers who make use of indirect ties to influence job assigning authorities are more likely to find better jobs. Belliveau, O'Reilly III, and Wade (1996) look at existing working relationships and find that social capital between a CEO and his or her compensation committee chairperson affects compensation decisions.

Other scholars do not focus on negotiation activities with (potential) employers, but use career success as their dependent variable. As Seibert, Kraimer, and Liden (2001) argue, the relationship between social capital and career success of individuals is fully mediated by three distinct types of network benefits. First, social capital is associated with information benefits. Second, social capital enables access to financial and material resources and, third, social capital is associated with legitimacy, visibility, and sponsorship. More recently, Cotton, Shen, and Livne-Tarandach (2011) examined the role of developmental networks of Major League Baseball Hall of Famers with respect to career success. Cotton et al. find that more successful Hall of Famers had a "broader range of developers across and within career communities" (2011: 25). As another example, social capital has been identifed as an important resource for immigrant entrepreneurs in the U.S. lodging industry as immigrant entrepreneurs are often supported by individuals from their ethnic group (Kalnins & Chung, 2006).

4.3.2 Group level studies

The social capital literature on group level focuses on both social capital within groups as well as across the boundaries of groups. With regard to intragroup social capital, Oh, Chung, and Labianca (2004) and Oh, Labianca, and Chung (2006) find that the relationship between group closure and group effectiveness follows an inverted U-shaped pattern. Also looking at intrateam networks, Balkundi, Kilduff, Barsness, and Michael (2007) find that moderate levels of structural holes are most conducive for team performance, while both very low and high levels of structural holes are detrimental for team performance. Addressing a similar research question, Balkundi and Harrison (2006) examine the influence of social capital on team performance and find that network density and leader centrality in an intrateam network are benficial for team performance[12]. Another scholar who looks at intragroup social capital is Gittelman (2007) who studies the role of regional as well as geographical dispersed ties in innovation project teams in the biotechnology industry. Gittelman (2007) finds that the results of projects composed of members from regional clusters are more likely to be cited in patents of the authoring firms, while geographically dispersed teams are more likely to produce results that are cited in scientific outlets.

With regard to external social capital of groups, Oh et al. (2006) suggest that a broad range of nonredundant relationships to other groups that are distributed among large numbers of group members are most benefical for group effectiveness. Oh et al. (2004) find that more relationships with formal leaders outside the focal group are associated with group effectiveness. Relatedly, Balkundi and Harrison (2006) suggest that centrality in interteam networks is benefical for team performance. Vissa and Chacar (2009) examine social capital of advice networks that cross entrepreneurial teams and find that the positive effect of structural holes in external advice networks is contingent upon two moderating aspects. First, strategic consensus and, second, network cohesion within the entrepreneurial teams affect the relationship between structural holes and venture performance positively. Hansen (2002) investigates cross-unit knowledge networks in the context of new product

[12] Note that the studies in this paragraph are further examples for the inconsistent findings with regard to the relationship between network structure and performance.

development projects. Hansen (2002) finds that the shorter a team's path lengths (i.e., the fewer intermediaries are necessary to connect another team), the higher is the amount of knowledge acquired and the shorter is completion time. However, the number of direct ties of a team is unfavorable if the knowledge transferred is codified as such circumstances do not necessarily require knowledge transfer via direct ties. Consequently, the cost for maintaining these ties are counterproductive. Stam and Elfring (2008) examine founding teams and ask how within industry network centrality and across industry bridging ties moderate the relationship between entrepreneurial orientation and firm performance. Concluding from their analysis, Stam and Elfring propose that "the configuration of a founding team's intra- and extraindustry social capital can explain both positive and negative performance effects of a firm's entrepreneurial orientation" (2008: 107).

4.3.3 Organizational level studies

A large number of scholars have shown that social capital affects organizational performance significantly (e.g., Gilsing, Nooteboom, Vanhaverbeke, Duysters, & Vandenoord, 2008). As the following quote of Florin, Lubatkin, and Schulze puts it: "[S]ocial resources leverage the productivity of a venture's resource base" (2003: 380). For instance, Cattani, Ferriani, Negro, and Perretti (2008) show how network ties between film producers and distributor organizations are beneficial for the survival of the former. More specifically, dense networks, which are characterized by repeated interactions and small turnover of the distributor organization, are associated with legitimacy, which is ultimately associated with firm survival. Fischer and Pollock (2004) find that social capital is a significant predictor of success in the context of initial public offerings. In more detail, both internal (i.e., the time managers worked together) as well as external (i.e., deal network embeddedness) social capital is negatively related to firm failure. As another example, Leana and Pil (2006) find that external as well as internal social capital of urban public schools is positively associated with organizational performance (i.e., student's learning achievements)[13]. Shipilov (2009) finds that sparse networks are especially conducive for companies

[13] More specifically, in the case of maths, the relationship between social capital and learning achievements is fully mediated by instruction quality of classes. In contrast, in the case of reading, social capital influences learning achievements directly (Leana & Pil, 2006).

that have a wide scope of experience (because these companies possess higher levels of absorptive capacity), for companies that are in multi-market-contacts with their peers (because it reduces the risk of non-cooperation), and for companies at the periphery of the focal network (because this position allows companies to contrast opportunities and information gained from different partners).

Some scholars have even examined a higher organizational level of social capital as they have investigated social capital between different clusters of companies[14]. For instance, Bae, Wezel, and Koo (2011) study ties that bridge geographically distant regional clusters of companies in the U.S. Biotech industry and find that such ties are beneficial for the formation of new companies. In their arguments, such cross cutting ties create access to information and other resources that enable entrepreneurs in the region to establish a business successfully. Eisingerich, Bell, and Tracey (2010) suggest that networks of regional clusters benefit from network strength and openness. Yet, while the benefits of network openness increase with environmental uncertainty, the benefits of network strength decrease.

4.3.4 Cross-level studies

Different levels of social capital do not exist independently, but are highly interrelated (Moliterno & Mahony, 2011). While some have highlighted the importance of further research in this domain (Moliterno & Mahony, 2011; Payne et al., 2011), a number of scholars have already worked on this issue.

For instance, extant research provides insights into effects of individual level social capital on organizational level outcomes. Kor and Sundaramurthy (2009) find that companies benefit from the social capital of outside directors that influence management as they monitor choices of board members. For example, companies that possess boards that are equiped with members who are involved in multiple boards grow faster. Acquaah (2007) finds that networking relationships on the individual level with managers at other firms, government officials as well as community leaders affect organizational performance positively. Furthermore, Acquaah (2007) examines how the positive influence of social relationships differs between

[14] Referring again to Moliterno and Mahony (2011), this meta-organizational analysis of organizational social capital is a result of a summary of a number of network members (i.e., organizations) to a higher level construct (i.e., cluster of organization).

organizational strategies. For instance, companies that pursue a low cost strategy do not benefit from relationships to managers of other companies, while companies that pursue a differentiation strategy do. Others have studied how organizational level social capital affects individual level performance. For example, Paruchuri (2010) examines how an individual's position in an intrafirm network interacts with structural centrality and structural holes on the organizational level to determine the impact of a focal individual on innovation activities. Paruchuri (2010) shows that the relationship between intrafirm network centrality and innovation involvement follows an inverted U-shaped pattern and the threshold of this relationship is affected by the position of the focal company in the interfirm network.

To sum up, extant scholars have demonstrated the importance of social capital across different levels of analysis. This multilevel characteristic of social capital is important to motivate the research gap I address in this dissertation. This aspect is discussed in the next chapter, where I present my research question.

5 Derivation of research gap and research question

The theoretical background presented within this part of the dissertation highlights both the rising relevance of innovation communities in research and practice (Dahlander & Frederiksen, 2012) as well as the importance of social capital in open innovation settings (Fey & Birkinshaw, 2005). In consequence, a number of scholars on open innovation (Nambisan & Baron, 2010; Rost, 2011) and innovation communities (Dahlander & Frederiksen, 2012; McLure Wasko & Faraj, 2005) have acknowledged social capital as a focal antecedent of performance. For instance, Grewal, Lilien, and Mallapragada (2006) find that network embeddedness of OSS projects positively influences technological project success. As another example, Björk and Magnusson (2009) examine an intraorganizational innovation community and find that while network centrality of an individual is positively related to the quality of submitted ideas, the relationship between network centrality of groups and idea quality follows an inverted U-shaped pattern.

While academic discourse has reached consensus that social capital helps to explain performance in innovation processes and innovation communities, extant research lacks with regard to antecedents of social capital. Payne et al. argue in their latest review on the topic that the majority of social capital researchers have studied performance implications of social capital while "there is a relative dearth of studies that analyze social capital antecedents" (2011: 500). Similarly, Sasovova, Mehra, Borgatti, and Schippers highlight the lack of research with regard to antecedents of social capital and state: "Research has tended to treat social networks as relatively static […]. But there is growing recognition that networks are in fact dynamic systems" (2010: 639).

The few studies that have examined antecedents of social capital in open innovation settings and innovation communities have limited their analysis to single dimensions or single levels of analysis. For instance, Fey and Birkinshaw (2005) find that partnering is superior to contracting with external partners in open innovation. These scholars implicitly propose that the governance mode of open innovation activities (i.e., contracting versus partnering) is an antecedent of relational social capital. Hence, Fey and Birkinshaw (2005) focus on antecedents of relational social capital on the organizational level. As another example, Dahlander and Wallin (2006) examine how an individual's sponsoring by a firm affects the structural dimension of innovation community social capital and find that firm sponsored community members form more social ties to other members of the community. Hence, Dahlander and Wallin (2006) restrict their analysis to antecedents of structural social capital on the individual level. The following statement of Carpenter and colleagues mirrors the general criticsm of extant researchers with regard to a too narrow focus on antecedents on the structural dimension of social capital: "Network development research focuses on the development and evolution of networks as social phenomena, mainly concentrating on patterns and precursors of network formation and change" (2012: 1340).

Notwithstanding the significant contribution of extant researchers on antecedents of social capital and innovation communities, one particular outstanding issue remains. A number of scholars have emphasized that different dimensions (Tsai, 2000) and levels of analysis (Moliterno & Mahony, 2011) of social capital are highly interrelated. For example, Tsai and Ghoshal (1998) find that the structural as well as the cognitive dimension of social capital antecede the relational dimension of social capital. With regard to the interrelations between different levels of analysis, Berends et al. (2011) describe how "contacts" on the individual level can result in "contracts" on the organizational level. Given the interrelations between multiple dimensions and multiple levels of social capital, I am concerned whether an analysis of antecedents of innovation community social capital that is restricted to one dimension or level is sufficient to gain comprehensive insights into the aspects that determine social capital in such settings.

To the best of my knowledge, there is no study that examines antecedents of social capital in innovation communities across multiple dimensions and levels of analysis. Within this dissertation, I consider multiple dimensions and levels of social capital and contribute to the closure of the research gap described in this chapter by asking:

What are antecedents of social capital in corporate innovation communities?

III

Antecedents of social capital in corporate innovation communities on three levels:

A case study approach

1 Structure of part III

Part III presents the first empirical study I have conducted in order to deal with the research gap presented in the previous part. I used a multiple case study approach (Eisenhardt, 1989) in order to contribute to the literature on antecedents of corporate innovation community social capital. My empirical setting is five innovation communities supported by different organizations of diverse sizes and with different industry backgrounds. Company size ranges from 500 to 360,000 employees. Industry settings vary from toy manufacturing to automobile. My primary data sources are 72 semistructured interviews with community members as well as community managers. As recommended by Jick (1979), I triangulated the interviews with other data sources, i.e., 14 community workshop observations, three group interviews, and analysis of interaction on community platforms.

I analyzed interviews with no fixed a priori coding scheme in mind. Yet, I considered the multidimensional and multilevel nature of social capital while analyzing and coding the interviews in order to build a comprehensive framework on antecedents of social capital in corporate innovation communities. The emergent theoretical framework comprises two important facets. First, the framework emphasizes the self-reinforcing nature of social capital and describes interrelations between different dimensions of the concept that lead to this self-reinforcing characteristic. Second, the framework categorizes antecedents of social capital on three different levels. Individual level antecedents capture those antecedents that affect single individuals. On this level, I have identified personality and motivation to engage in community work as central antecedents of community social capital. Organizational level antecedents entail aspects that affect all members of the organization that supports community activities, irrelevant of community membership. This level of antecedents is relevant in the context of my dissertation as the supporting organization provides the environment upon which social capital in community settings develops. In essence, according to the definition of corporate innovation communities, some community members are part of the supporting organization. On the organizational level, I identified corporate culture as well as formal and informal organization as antecedents of innovation community social capital. Finally, community level antecedents are those antecedents that affect all members of the focal corporate innovation community. Antecedents that belong to this category include creative task design, community leadership as well as workshop and platform design.

Part III is organized based on Eisenhardt's (1989) suggestions. *First,* I discuss general methodological issues of case study research and why it is appropriate to tackle the research question I aim to answer. In more detail, I stress the exploratory nature of the research presented, the relevance of contextualism and multiple levels of analysis, as well as the processual nature of social capital development. *Second,* I present my case selection procedure. I discuss advantages and disadvantages of single and multiple case designs, and present my procedure of analysis of a longitudinal pilot case followed by four additional cases. This approach allowed me to combine benefits of both single as well as multiple case designs. In addition, I stress the importance of theoretical sampling in case study research and provide detailed information on the five cases studied. All cases under research are similar in a sense that they fit the definition of a corporate innovation community I have presented in part II of this dissertation. Yet, they differ to some extent, most importantly with regard to three aspects: (a) Industry affiliation of the cases presented varies from toy manufacturing to automobile or temporary employment; (b) my sample includes different company sizes ranging from 500 to 360,000 employees; (c) while some cases under research are corporate innovation communities that operate within the boundaries of companies (n=2), others consist of innovation communities that are composed of internal as well as external community members (n=3). *Third,* I turn to issues of data collection and analysis. This chapter provides information on the number of interviews analyzed in each case, interviewee characteristics, interview guidelines used as well as additional data sources (i.e., workshop observations and interaction on online platforms) that allowed me to triangulate my findings. In total, I analysed 72 interviews, oberserved 14 community workshops and analysed interaction on online platforms of three communities. Furthermore, I describe how data collection and analysis overlaped which allowed me to adapt data collection procedures within the research process. At the end of the third chapter, I describe my approach with regard to within as well as cross case analysis. *Fourth,* I present the theoretical framework developed in the context of this study, which entails two important facets. First, the framework conceptualizes social capital as a self-reinforcing phenomenon, which is composed of three interrelated dimensions. Second, the framework offers antecedents of corporate innovation community social capital on individual, organizational as well as community level. *Finally,* in the discussion and conclusion chapter, I discuss contributions of this study to the literature as well as aspects of my findings that differ and contradict extant publications. I end with limitations and avenues for future research. Managerial implications of the study are discussed in part V of this dissertation. Figure 10 depicts the structure of part III.

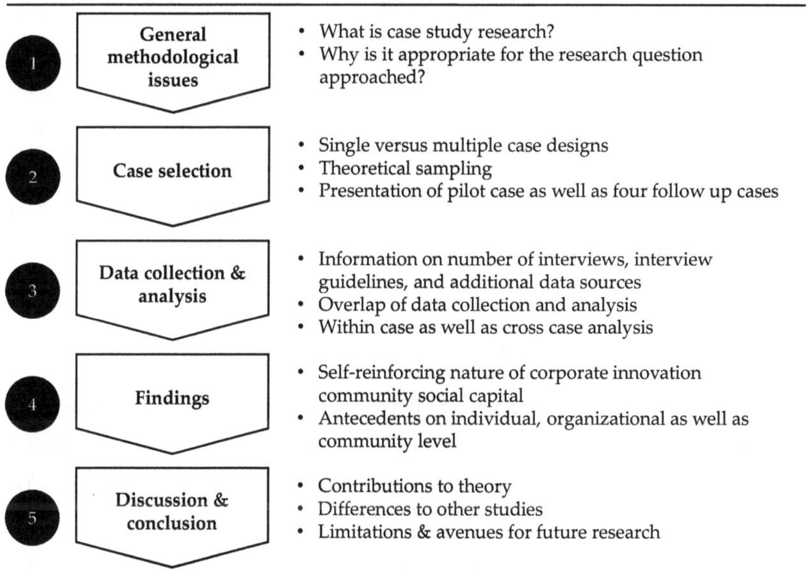

Figure 10: Structure of part III based on Eisenhardt (1989)

2 General methodological issues

Given the limited scientific understanding of the phenomenon under research in this part of the dissertation, I use a multiple case study approach as proposed by Eisenhardt (1989) and (Yin, 2009)[15]. My goal is to "develop pertinent hypotheses and propositions [of antecedents of corporate innovation community social capital] for further inquiry" (Yin, 2009: 9). Yin defines a case study approach as follows: "[T]he distinguishing characteristic of the case study is that it attempts to examine: (a) a contemporary phenomenon in its real-life context, especially when (b) the boundaries between phenomenon and context are not clearly evident" (1981: 59). The advantages of case study research as an analysis of real-life contexts are also highlighted in another, often cited work of Yin: "[T]he distinctive need for case studies arises out of the desire to understand complex social phenomena. In brief, the case study method allows investigators to retain the holistic and meaningful characteristics of real-life events" (2009: 4).

Case study research occupies the "ideal of no theory under consideration and no hypotheses to test" (Eisenhardt, 1989: 536). In this context, extant scholars emphasize both the importance of an adequate research question (Yin, 2009) as well as specification of relevant constructs (Eisenhardt, 1989) prior to case study data collection and analysis. First, an adequate research question enables scholars to handle the complexity of the real-life situations observed in case study settings (Yin, 1981) and forestalls "death by data asphyxiation - the slow and inexorable sinking into the swimming pool [of data] which started so cool, clear and inviting and now has become a clinging mass of maple syrup" (Pettigrew, 1990: 281). At the same time, Yin (2009) notes that the research question should be broad enough to preserve an exploratory research approach. Second and relatedly, Eisenhardt (1989) stresses the importance to give thought to constructs that are likely to be relevant within the

[15] I am aware that Eisenhardt (1989) and Yin (2009) differ in their perceiptions of the inductive versus deductive nature of case study research. While Eisenhardt's (1989) approach is closely related to the notion of grounded theory (Glaser & Strauss, 1967) and inductive research, Yin (1981) suggests that case studies share numerous similarities with experiments and deductive research. I follow the arguments of Pettigrew (1990) who suggests that case studies entail both inductive as well as deductive elements.

research question approached. Yet, Eisenhardt (1989) also states that there is no guarantee in advance that any construct will find its place in the resultant theoretical framework.

I propose that my research question is both broad enough to ensure exploratory research efforts as well as specific enough to avoid data asphyxiation. With regard to inductive research, my aim is to get initial insights into antecedents of social capital in corporate innovation communities - a phenomenon that has not received enough attention in the literature so far. This is surprising because a number of scholars have highlighted the importance of social capital in community settings (Dahlander & Frederiksen, 2012; McLure Wasko & Faraj, 2005). Furthermore, research on antecedents of social capital contributes to knowledge on managerial levers with regard to community networks and, hence, is highly relevant for practitioners. In this context, my aim is not to test theoretically developed relationships, but to generate a theoretical framework, which presents propositions that offer avenues for further confirmative studies. With respect to relevant constructs, I take care of both multiple dimensions (Nahapiet & Ghoshal, 1998) as well as multiple levels (Payne et al., 2011) of social capital. This perspective allows me to draw a holistic picture of antecedents of corporate innovation community social capital.

I argue that a case study approach is particularly appropriate in the context of this part of my dissertation, most importantly because of two underlying aspects over and above the exploratory characteristic of this research. First, case studies are appropriate to advance scientific understanding of phenomenona of processual nature (Ozcan & Eisenhardt, 2009). My research tackles antecedents of social capital, whereby investigating the establishment of social relationships over time and, hence, clearly explores a processual phenomenon. Second, researchers have argued that a focal advantage of case study research is its capability to investigate multiple levels of analysis at the same time (Bourgeois & Eisenhardt, 1988). This issue is closely related to Yin's (1981) argument for the advantages of case studies in settings where boundaries between phenomena and context are unknown. As Pettigrew puts it: "Our [...] assumption of contextualism is that target changes should be studied in the context of changes at other levels of analysis" (1990: 269). Transferred to the research presented here, I propose that studies on antecedents of innovation community social capital need to take a look at deeper (i.e., individual) and higher (i.e., organizational) levels of analysis in addition to the community level. Figure 11 depicts my arguments for the appropriateness of case study methodology in this study.

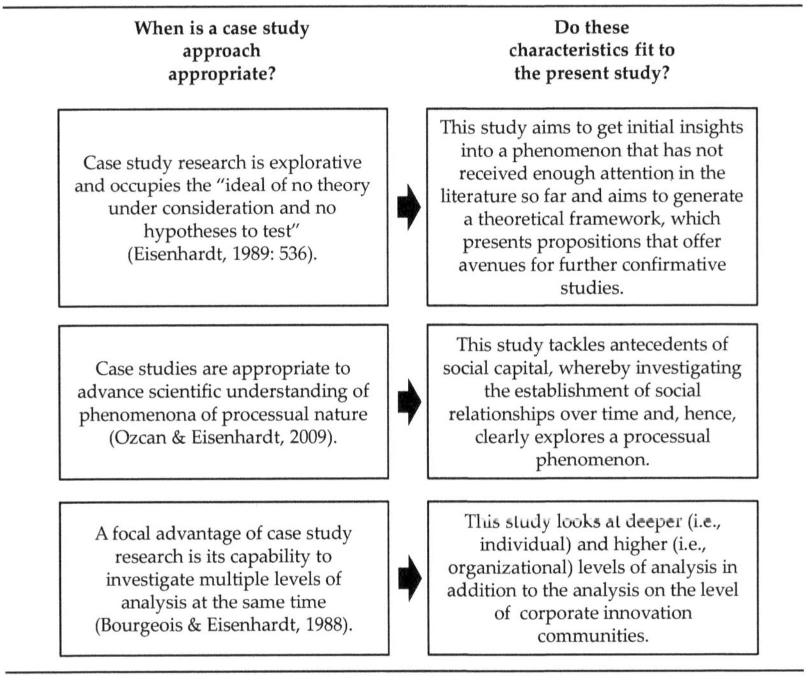

Figure 11: Case study methodology and characteristics of the present study

Related to that, a number of scholars have relied on case studies as a means to advance scientific understanding of social capital and networks. For instance, Gulati and Puranam (2009) investigate at Cisco how formal organization and informal networks interact. Observing a reorganization process, Gulati and Puranam (2009) find that Cisco benefits from a survival of old organizational structures in informal networks and achieves the capability of ambidexterity. As another example, Mosey and Wright (2007) use a multiple case study approach to study how academic entrepreneurs proceed as they establish their business. One major finding of Mosey and Wright (2007) is that academic entrepreneurs benefit from prior business ownership because these entrepreneurs are able to develop new network ties effectively.

3 Case selection

Similar to other research methods, selection of the empirical basis represents a crucial step in the context of case study research. A central discussion in this context addresses benefits and drawbacks of single versus multiple case study designs. Those who argue for benefits of single case studies often emphasize the advantages to gain deep and rich insights into processes under research (Bergenholtz, 2011; Dyer & Wilkins, 1991) and to observe the same case at different points of time in a longitudinal manner (Yin, 2009). A number of scholars in the field of social capital have referred to this argumentation and such single firm approaches are not uncommon in this research strand (e.g., Harrison & Waluszewski, 2008; Hughes & Perrons, 2011).

To the contrary, others argue that multiple case studies are superior to single case study designs (Martin & Eisenhardt, 2010; Ozcan & Eisenhardt, 2009). For instance, Yin (2009) criticizes shortcomings of single cases with regard to generalizability of findings and suggests that multiple cases allow more powerful data analysis. More specifically, additional cases allow a researcher to replicate the first case in other, different settings. As Eisenhardt puts it: "In replication logic, cases which confirm emergent relationships enhance confidence in the validity of the relationships. Cases which disconfirm the relationships often can provide an opportunity to refine and extend the theory" (1989: 542). Following this line of argument, a number of social capital scholars have relied on multiple cases in their research endeavours (e.g., Eisingerich et al., 2010; Maurer & Ebers, 2006).

In this part, I combine advantages of both single as well as multiple case studies as I rely on longitudinal data collection at a pilot case and proceed with four follow up cases. This procedure is recommended by Yin (2009) and allows me both to gain rich and deep insights in the longitudinal pilot case setting as well as to replicate initial findings as I analyse four additional follow up cases. In doing so, I am able to refine and to extend the theoretical model proposed. Figure 12 depicts the research design applied.

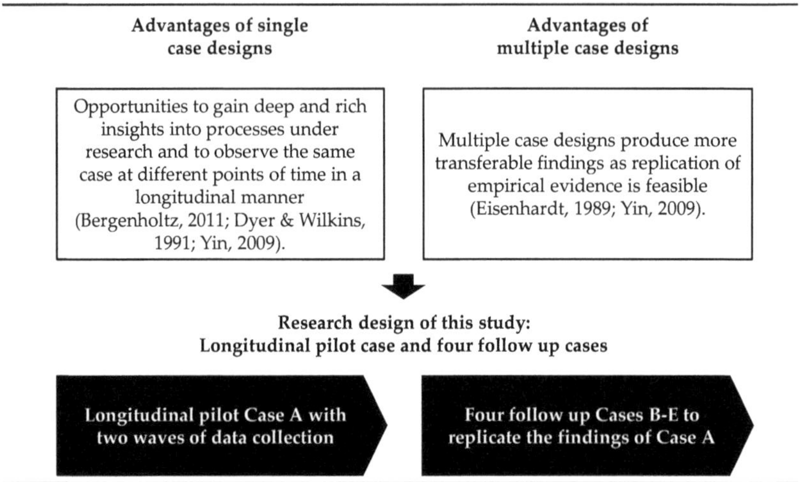

Figure 12: Advantages of single versus multiple case studies and research design

In selecting the case studies, I followed theoretical sampling as suggested by Eisenhardt (1989). In other words, I selected cases that allowed me to analyse antecedents of corporate innovation community social capital in a transparent way (Pettigrew, 1990). Furthermore, the cases selected allowed both literal as well as theoretical replication (Yin, 2009). All five cases I have investigated meet the definition of corporate innovation communities I have presented in part II and, as such, are suitable to produce transferable results. On the other hand, the cases differ with regard to industry characteristics, sizes, and detailed innovation community design and thus, my sample allows me to generalize findings across heterogeneous settings. I present the pilot case as well as the follow up cases in the following.

3.1 Pilot Case A: Temporary employment

The empirical setting of the pilot case (Case A) is the temporary employment industry in Germany. This setting is interesting for several reasons in the context of this dissertation. First, the temporary employment industry is part of the service sector, which represents an increasingly important economic factor. For instance, Bitner, Brown, and Meuter (2000) argue that, as the worldwide economy will become more service oriented, service innovation will gain on importance with respect to competitive advantages of organizations. Second, various researchers recently have begun to study open innovation facets in service industries (Carbonell, Rodríguez-Escudero, & Pujari, 2009; Oliveira & von Hippel, 2011). Third, competitive conditions within the temporary employment industry make this empirical field interesting with regard to innovation processes. As many organizations have increasingly relied on external workers in the recent past, the temporary employment industry has experienced a tremendous growth in the last decade (Davis-Blake & Uzzi, 1993). Furthermore, temporary employment agencies enable organizations to react flexible, for instance with regard to incoming orders (von Hippel, Mangum, Greenberger, Henemann, & Skoglind, 1997). As a result, the temporary employment industry faces a very dynamic environment and innovative capabilities are especially important.

Data has been collected at an innovation community supported by an organization belonging to the ten biggest companies within the focal industry in Germany. With more than 60 branches spread over Germany, social relationships between the 500 employees of the company are a crucial success factor for innovative performance. In order to further foster innovation processes across the boundaries of branches, the management decided to support the establishment of an innovation community by provision of offline workshops and an online innovation platform that allowed members of the community to exchange ideas. The management of the company decided not to open the innovation community to external members. In other words, innovation community members of Case A are employees of the focal company.

3.2 Four follow up cases

In this section, I present the four follow up cases I have analyzed in order to approach my research question. Information provided includes industry characteristics, company specifica, and facets of the corporate innovation community analysed.

3.2.1 Case B: Automobile

The empirical setting of Case B is the automobile industry. Tabbed by increasing price pressures, mature markets in industrialized countries as well as strengthened environmental and safety regulation (Ili et al., 2010), the need for innovative product and process innovations in the auto industry increases tremendously. Related to that, Rothenberg and Ettlie (2011) propose that automotive companies face both extraordinary market and regulatory uncertainty. These scholars describe how organizations in the automotive industry respond to these uncertainties with platform and functional integration in order to foster incremental as well as radical innovation. A particular prevalent facet for companies situated in the automobile industry is the transformation associated with rising gas prices and ecological awareness. For instance, Vandermey (2012) discusses the rising relevance of other fuels such as electronics or natural gases. Closely related to the issues I have discussed above, automotive companies have started to search for new ways to cope with future challenges and turn to more open paradigms of innovation as they cooperate with customers, suppliers, universities or communities in order to develop new products or services (Ili et al., 2010). For example, Gassmann and Enkel (2004) describe how European automotive companies benefit from mutual cooperation with an Austrian supplier in innovation processes. With regard to innovation communities, Füller and colleagues (2008) find that creativity, identification with the community, emotional and attitudinal attachments as well as brand knowledge affect the willingness of German automobile brand community members to engage in innovation processes positively.

The company analyzed employs about 60,000 people in different branches in Germany and Europe and cooperates with a number of external partners (e.g., universities and customers) in innovation processes. Furthermore, as knowledge and idea exchange across the boundaries of functions as well as branches are pivotal for the company to remain competitive, the company supports activities of an internal innovation community. In more detail, the management of the company is aware of the pivotal role of informal knowledge exchange between employees and supports such processes via offline meetings and online platforms. The innovation community investigated in Case B included only employees of the focal company. Individuals who do not belong to the focal company have no access to community activities.

3.2.2 Case C: Toy manufacturing

The company that supports the innovation community studied in Case C operates in the toy manufacturing industry. The toy industry is particularly interesting in the context of my dissertation because companies that compete within this industry experience increasing competition, for instance because of shorter product life cycles (Chung & Ng, 2011). Therefore, toy manufacturers encounter high pressures to develop innovative products. In essence, the toy industry nowadays develops innovative products and technologies which are also useful for other industries (Weintraub, 2002). For instance, a company that traditionally developed computer games is now responsible for the development of military training tools for the American Army (Weintraub, 2002). Beside the challenges associated with shorter product lifecycles, toy manufacturers have to deal with vigorous changes in sales markets. While the industry was traditionally dominated by sales via retail businesses, customers nowadays increasingly rely on online shops (Datamonitor, 2010). Other forces that are positively related to the need for innovation in the industry are global competition as well as the increasing relevance of digital technologies (Schultz & Hatch, 2003). In reaction, toy manufacturers have started to establish boundary crossing innovation processes that aim to create and capture value. For instance, the toy manufacturer LEGO heavily relies on online communities in order to integrate customers into innovation processes (Frey & Lüthje, 2011; Schultz & Hatch, 2003).

The privately owned company under research in Case C is located in Germany and employs approximately 2,000 employees. Every year, the company brings an impressive number of new products to market and has won several awards for its innovative toys. The management of the company is aware of the pressure for innovation, in particular because the cost structures of German toy manufacturers are significantly above those of other countries and customers are progressively interested in low cost products (Brill, 2012). As a result, the management supports innovation processes in multiple ways – for instance via collaboration with universities, suppliers, or customers. In addition, the company supports activities of an innovation community where both employees of the company as well as external community members innovate together to help the company to remain competitive.

3.2.3 Case D: Automation

A company that offers automation solutions supports the corporate innovation community that provides the empirical setting of Case D. The automation industry is interesting because of several reasons. Automation solutions combine new technologies in different fields such as computer science and robotics and have heavily changed the way how companies organize manufacturing processes (Kozlowski, 1987). As such, automation companies offer insights into core processes of technological innovations. The significant relevance of innovative automation technologies is documented in the study of Kelley (1994) who shows that automation technology reduces production time per unit tremendously. Strongly affected by emerging technological opportunities, companies that operate in the industry are constantly forced to innovate and keep pace with regard to new products and services (Bond, 2006).

The company that hosts the innovation community investigated employs about 15,500 employees worldwide. As a leading supplier of automation technology, the company invests an impressive amount of its turnover in R&D activities and launches about 100 product innovations every year. The company draws on both internal and external partners in innovation processes. Amongst many other initiatives, the management supports the work of innovation communities within as well as outside the boundaries of the organization.

3.2.4 Case E: Conglomerate

My last Case E is a community supported by a conglomerate, which operates in multiple industries and countries. Such companies play a pivotal role in the context of innovation, for instance because these conglomerates are able to provide emerging technologies legitimacy (Lange, Boivie, & Henderson, 2009). As Hitt, Hoskisson, and Kim (1997) argue, international and product diversification affect a company's innovativeness in an interrelated and complex manner. In more detail, international diversification is positively related to innovation because it enables companies to benefit from selectivity advantages in multiple countries. Furthermore, conglomerates often have the necessary resources at hand to invest in R&D and are motivated to do so because these conglomerates are able to capture the value of resulting innovations in multiple countries. To the contrary, Hitt et al. (1997) argue that product diversification is detrimental to innovation because it is associated with information and knowledge asymmetries between different product lines. Related to that, the literature provides a number of studies that have emphasized the importance of social capital for innovative performance of conglomerates (e.g., Tiwana, 2008) as social capital can promote knowledge exchange between different units (Tsai, 2001).

The company that supports the community examined employs approximately 360,000 people worldwide. Its strong focus on technological innovation is observed by taking a closer look at the organization: Around 30,000 people work in R&D departments of the company. The organization cooperates with various external partners such as customers and universities in innovation processes. In this context, the company supports innovation community activities both within as well as across the boundaries of the organization by provision of online as well as offline tools.

Table 2: Overview of cases

Case	*Industry*	*Size of supporting organization*	*Community design*
A	Temporary employment	500 employees	All community members are employed by the supporting organization.
B	Automobile	60,000 employees	All community members are employed by the supporting organization.
C	Toy manufacturing	2,000 employees	Community members are employees as well as external individuals.
D	Automation	15,500 employees	Community members are employees as well as external individuals.
E	Conglomerate	360,000 employees	Community members are employees as well as external individuals.

4 Data collection and analysis

In this chapter, I present methodological aspects related to data collection[16] and analysis. I present data sources within each case including detailed information with regard to interviewee characteristics. The data collection approach differs between the pilot Case A and the four follow up Cases B-E. In consequence, I present the longitudinal data collection approach of the pilot case first and continue with the data collection at Cases B-E. I end with a description of data analysis procedures.

4.1 Data collection: Case A

I used multiple data sources at Case A: Interviews, observations at community workshops, and log files that allowed me to analyse interaction on an online platform where community members exchanged ideas. The triangulation of different sources allowed me to obtain a richer set of information and to increase the robustness of my developed theory (Anand, Gardner, & Morris, 2007). As Jick put it: "Given basic principles of geometry, multiple viewpoints allow for greater accuracy. Similarly, organizational researchers can improve the accuracy of their judgments by collecting different kinds of data bearing on the same phenomenon" (1979: 602).

The primary data source to analyze the corporate innovation community representing Case A was interviews with community members and employees of the company that supported community activities. In total, 30 interviews were conducted, which relied on open ended questions during most parts. A key advantage of the longitudinal research is the two waves of data collection, beginning in November 2010. A further advantage is that data could be gathered before the management began to support community activities. As a result, it was possible to identify the specific effects of community aspects that affected social capital.

The first wave of interviews took place before the management started to support community activities. Respondents from multiple hierarchical levels (ranging

[16] Other researchers have been involved in the process of data collection as well. See annex 1 for further information.

from apprentices to branch managers) and from different regions were asked about innovation processes as well as knowledge and idea exchange. Interviews started by asking respondents to report the implementation of a successful innovation at the company. Afterwards, questions addressed innovation processes in general (e.g., "How does the company search for good ideas?", "How does the company evaluate ideas?", "How does the company implement ideas?") as well as knowledge and idea exchange between employees during such innovation processes (e.g., "Do you regularly exchange ideas with your colleagues?", "Do you often talk to employees from other branches in order to exchange ideas?", "Does the management support exchange between employees from different branches?"). During this wave of data collection, nine employees have been interviewed in face-to-face settings and 12 employees have been interviewed via telephone. Most interviewees worked in very common functions, i.e., sales (five interviewees) and branch administration (ten interviewees). In addition, four branch managers and two apprentices have been interviewed in order to get insights into different hierarchical levels. Interviewees worked at 17 different branches across Germany. Interview length varied from 24 to 51 minutes, with an average of 35 minutes.

The second wave of interviews took place in August 2011, nine months after the first wave. In the meantime, the management had organized workshops and launched an online platform in order to support community activities. Interviews started with closed-ended questions concerning the strength of respondents' ties to other community members before community activities have been supported. On this basis, respondents were asked to describe how network structures, trust, and shared visions and languages among community members had changed since then. The guideline included questions concerning all dimensions of social capital, i.e., the structural, the relational, as well as the cognitive dimension. Again, respondents from several hierarchical levels as well as from several regions were interviewed. Nine interviews via telephone with community members from seven different branches as well as the headquarter of the company were conducted. The sample included one employee who was responsible for the management of a whole region, which included several branches, two branch managers as well as three sales employees and one employee responsible for administrative tasks within branches. Two of the interviewees worked at the headquarter of the company which supports the branches, for instance via a central key account management. Apart from their daily duties according to their job descriptions, interviewees in the second wave participated at the corporate innovation community on a voluntary basis. The duration of the interviews varied from 18 to 35 minutes (average duration: 27 minutes).

Potential informant biases of interviewees in both waves were addressed in different ways. First, employees from several hierarchical levels have been interviewed. For example, some interviewees were apprentices, while others were managers responsible for several branches. Second, interviewees worked at multiple branches located at different regions in Germany. Some worked at a branch in Hamburg in the north of Germany and others worked in Bavaria in the south of Germany. Third, in order to avoid recall biases, questions addressed recent accounts. Because of the subjective nature of evaluations of social relationships, an important advantage of the methodological approach was the ability to study the establishment of the innovation community from its very beginning (Golden, 1992). Fourth, as mentioned above, I triangulated data from multiple sources in order to reassess the information obtained during the interviews (Jick, 1979). Apart from the interviews, I relied on two secondary data sources in this study. In more detail, I was able to observe two community workshops and analyzed interaction of community members on an online platform. Pettigrew (1990) recommends the use of different types of data such as interviews and observations because data sources have specific strengths and weaknesses. For instance, while interviews allow detailed analysis of personal feelings, observations provide "access to group processes and can confront the researcher with discrepancies between what people have said in interview and casual conversations, and what they actually do" (Pettigrew, 1990: 277). Fifth, anonymity was assured to the interviewees (Martin & Eisenhardt, 2010). Finally, interviewees were highly interested in community activities. As a result, interviewees could be convinced that this research is of importance, which allowed accurate information gathering.

4.2 Data collection: Case B-E

In this section, I present the data collection approach with respect to the four follow up Cases B-E. In contrast to Case A interviews, I did not conduct all interviews on my own. In some cases, other scholars who have been informed about my research conducted the interviews. Most of the interviews were conducted by two researchers. This procedure is recommended by Bourgeois and Eisenhardt, who argue for the advantages of tandem interviews "with one investigator primarily responsible for the interview and the other responsible for taking notes and filling in gaps in the questioning" (1988: 819).

Another important aspect of my research concerns an overlap of data collection and data analysis (Eisenhardt, 1989). In other words, I started with data analysis of the pilot case before data at the four follow up cases was collected. This process allowed me to adapt and extend data collection instruments. Faems, Janssens, and van Looy describe a similar procedure as they experience an "evolution of surface levels to deeper levels of data collection and analysis" (2010: 8). I describe the data collection for Case B as well as Case C-E, where similar interview guidelines were used, in the following.

4.2.1 Data collection: Case B

The automotive company that supports community activities analyzed in Case B is very experienced in the management of innovation communities. In stark contrast to my pilot case, community-based innovation processes were common in the organization for many years. My interest in Case B focused on the organizational processes that support innovation communities. Opposing to Case A, my interest shifted from community members to those individuals who shape community processes from an organizational point of view. In this context, three groups of people have been identified as a promising source of information: (a) R&D management, (b) employees responsible for strategy, organization, and corporate culture (i.e., general management), and (c) knowledge and idea management. With respect to the first group, interviews with two individuals were conducted, one responsible for the organization of R&D processes and one whose task is the support of these processes via information technology (IT). Furthermore, I analyzed interviews with an inhouse consultant, with a person who was responsible for coaching and human resource (HR) development, with a manager responsible for process and project organization, and an employee affiliated to the strategy department, representing the second group of interviewees. Finally, to cover the third group of interest, I analyzed interviews with a person responsible for the idea management as well as a person who manages the social media that is used within the boundaries of the company. In total, I analyzed eight interviews which varied in duration from 45 to 92 minutes and averaged at 67 minutes. All interviews were conducted in face-to-face settings and took place in early 2012.

The semi-structured interview guideline used at all eight interviews started with a short self-introduction of the interviewee followed by aspects of corporate culture and leadership related to innovation community processes (e.g., "Please describe your organization's corporate culture in three sentences!"). In the next section of the interviews, the guideline proposed questions with regard to knowledge exchange and social capital in the organization (e.g., "Please describe the role of social relationships within your organization!"). Finally, interviewees were asked to describe and discuss offline as well as online instruments that supported social capital and community-based innovation processes (e.g., "What technologies does your company use to support innovation community activities?"). Potential informant biases were addressed as employees from several hierarchical levels and branches have been interviewed. Furthermore, the questions in the interview guideline addressed recent accounts wherever possible (Golden, 1992). For instance, interviewees were asked to compare the actual corporate culture with the corporate culture at earlier times. Finally, anonymity to the interviewees was assured (Martin & Eisenhardt, 2010). Unfortunately, I had no access to additional data sources.

4.2.2 Data collection: Case C-E

In this section, I discuss the data collection approach at the companies that support innovation communities in the context of Cases C-E. I summarize these three cases in one section because the data collection approach in all three corporate innovation communities was similiar. All three companies decided to support innovation communities that foster knowledge and idea exchange across functional and organizational boundaries. In addition, the companies wanted to recruit innovation community members of different ages in order to integrate different perspectives in innovation community activities.

I analyzed interviews with managers of the companies as well as with community members. The interview guideline used for managers started with a short self-introduction of the interviewee followed by questions that address general organizational issues including facets of social capital (e.g., "How do you support social relationships between employees?") as well as innovation management of the focal organization (e.g., "How do you organize new product development in your company?"). The guideline ended with questions regarding the supported innovation community (e.g., "What barriers do exist in your organization that are detrimental to community participation of employees?"; "Would you describe social ties within the community as trustful?").

The interview guideline for the community members also started with a short self-introduction of the interviewee. Apart from the self-introduction, the guideline contained three parts. First, the guideline included questions addressing general information on the organization including social capital (e.g., "Please describe the role of social relationships at your company!"; "Would you describe your relationships with your colleagues as trustful?"). The following sections addressed aspects of innovation management at the company (e.g., "What opportunities do you have in your organization to come up with innovative ideas?") as well as the innovation community in particular (e.g., "Why do you participate in community activities?"; "Do community members share a common vision of their work?"). I analyzed five interviews with the management and eight interviews with community members at the toy manufacturing company. Seven community members were part of the organization and one member was external. Ten interviews were conducted in face-to-face settings and three interviews were conducted via telephone. The interviews varied from 8 to 91 minutes and averaged at 49 minutes.

At Case D, I analyzed two interviews with the management (one employee affiliated to the HR department and one employee working for the technology and innovation management department). In addition, I analyzed five interviews with community members who were employed by the focal organization and three interviews with community members who were not employed by the focal organization. Interview length varied from 50 minutes to 83 minutes with an average at 67 minutes.

Finally, I analyzed nine interviews with internal and two interviews with external community members at the conglomerate (Case E) (variation in length from 61 minutes to 86 minutes; average: 74 minutes). Interviews at Case C took place in winter 2011/2012, while interviews at Case D were conducted in autumn 2012, and interviews at Case E took place in summer 2012.

In addition to the interviews, I triangulated data with 12 workshop observations (six at the community supported by the toy manufacturer (Case C) and six at the community supported by the company that offers automation technologies (Case D)) as well as analysis of interaction on online community platforms at the toy manufacturer (Case C) and the conglomerate (Case E). Potential informant biases were addressed by choosing interviewees from different hierarchical levels and recall biases were avoided by addressing recent accounts (Golden, 1992). Interviewees were asked to report on community activities that took place in the last months. In addition, the triangulation of multiple data sources (Jick, 1979) as well as assured anonymity to the interviewees (Martin & Eisenhardt, 2010) worked against potential informant biases. The data collection approach is summarized in Table 3. In the next section, I present my data analysis procedure.

Table 3: Overview of interviews and other data sources

Case (Industry)	*# of interviews*	*Interviewee characteristics*	*Duration of interviews*	*Additional data sources*
Case A (Temporary employment)	Two waves of data collection with 21 and 9 interviews	Employees and community members from diverse branches, functions, and hierarchical levels	First wave: 24 to 51 minutes with an average at 35 minutes; Second wave: 18 to 35 minutes with an average at 27 minutes	2 workshop observations; Interaction on IT-based platform
Case B (Automobile)	8 interviews	R&D management, general management, as well as knowledge and idea management	45 to 92 minutes with an average at 67 minutes	None
Case C (Toy manufacturing)	13 interviews	Management as well as community members	8 to 91 minutes with an average at 49 minutes	6 workshop observations; Group discussion; Interaction on IT-based platform
Case D (Automation)	10 interviews	Management as well as community members	50 to 83 minutes with an average at 67 minutes	6 workshop observations; Group discussion
Case E (Conglomerate)	11 interviews	Community members	61 to 86 minutes with an average at 74 minutes	Group discussion; Interaction on IT-based platform

4.3 Data analysis

My data analysis approach was characterized by constant and repeated comparison of empirical information and theory (Eisenhardt, 1989). As I have described above, I started with the analysis of data gathered at the community supported by the temporary employment company (Case A). I followed the recommendation of Yin (1981) who argues for a preliminary data analysis which aims to gain initial understanding of the case and precedes any formal coding activities. Hence, I started by studying interview data and workshop observation notes without assigning specific codes or categories to gain an initial toehold with respect to the focal case. Afterwards, I coded the transcripted interviews.

It is important to note that I had no a priori coding scheme. In contrast, my coding scheme emerged from data as I analyzed the interviews separately. Yet, while coding the interviews, I had the multiple dimensions and multiple levels of social capital in mind. After the first coding, my approach followed an iterative cycling between multiple data sources and theory in order to advance scientific understanding with regard to my research question. If an interviewee mentioned a factor that antecedes social capital, I checked antecedents already mentioned before and tried to summarize these antecedents in categories. If a factor mentioned was not captured by any of my extant categories, I opened a new category. The procedure applied is closely related to the work of Glaser and Strauss (1967). Bouty (2000) provides a useful summary of this research approach. As I discuss in the findings section of this part, two important insights emerged. First, social capital is of self-reinforcing nature, i.e., the different dimensions of social capital are highly interrelated. Second, I structure antecedents of community social capital on three different levels (i.e., individual, organizational, as well as community level). As mentioned, I triangulated interview data with other sources to increase the robustness of my findings. In addition, I followed an iterative approach, cycling back and forth from data to theory and vice versa until I achieved a satisfying fit between data and my theoretical framework.

As I have mentioned, there was a strong overlap between data collection and analysis in my study. Data collection at communities representing my Cases B-E started after I had completed a preliminary analysis of Case A (Yin, 2009). Based on initial insights, data collection procedures (e.g., interview guidelines) were adapted as suggested by Eisenhardt (1989) in order to benefit from deeper understanding. My first step was detailed within case analysis. My aim in this context was to "become intimately familiar with each case as a stand-alone entity" (Eisenhardt, 1989: 540) before I moved forward to generalize findings across cases. This within case analysis was an important step to avoid the danger of "death by data asphyxiation" (Pettigrew, 1990: 281). In this context, each single case allowed me to replicate the findings and to derive more robust and valid insights (Yin, 2009). This within case analysis was followed by a cross case analysis, in other words a comparison of the findings between the five different cases (Eisenhardt, 1989). As others have proposed in the past (e.g., Harris & Sutton, 1986), my aim in this context was to generate a framework that is transferable to other settings. I present my findings in the next section.

5 Findings on antecedents of corporate innovation community social capital

In this section, I present my inductively derived framework, which offers theoretical logic to understand factors that influence the three dimensions of social capital in corporate innovation communities. My sample includes both corporate innovation communities that are restricted to members of the supporting organization (Cases A and B) as well as communities that also integrate external members (Cases C-E).

Some of the arguments I use to provide a theoretical basis for my model originate from the literature on innovation communities that appear self-organized (Franke & Shah, 2003) and are not supported by companies. While some of my findings are also applicable to self-organized innovation communities, most facets of my framework address issues that are restricted to company supported innovation communities.

As depicted in Figure 13, the framework presented in this chapter is composed of two facets. First, it describes social capital as a self-reinforcing asset as different dimensions of the concept are highly interrelated and amplify each other. Second, the self-reinforcing, multidimensional concept of social capital is determined by three interrelated levels of antecedents of innovation community social capital. I propose that antecedents of community social capital reside at the individual, the organizational, and the community level. Individual level antecedents contain aspects directly related to single community members. In contrast, organizational level antecedents refer to facets affecting the organization that supports community activities as a whole. Finally, community level antecedents enclose aspects that affect the community as a collective.

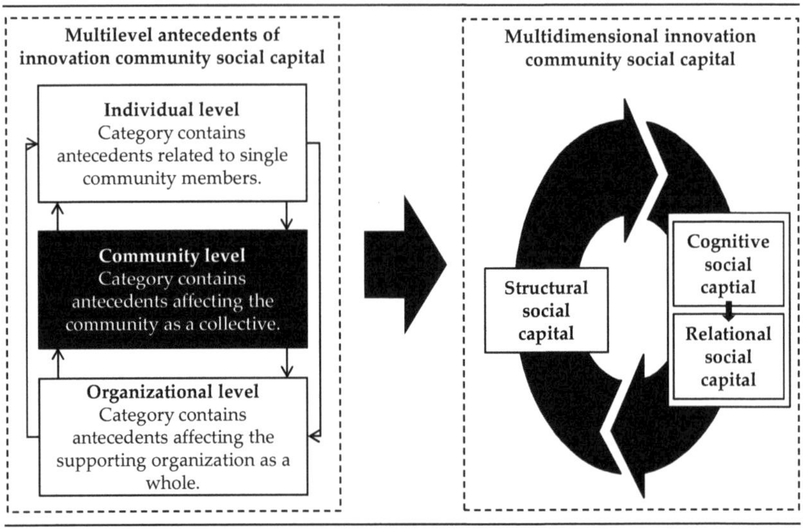

Figure 13: Structure of the proposed framework

My categorization of antecedents of corporate innovation community social capital highlights that there are two important categories of antecedents over and above those that are situated at the community level. Referring to the notion of contextualism (Pettigrew, 1990), I argue that individual and organizational levels of analysis are relevant, too. First, because corporate innovation communities consist of numerous members, antecedents on the individual level are also relevant in determining social capital in such settings. Second, corporate innovation communities receive support by an organization that provides some of its employees as community members. As a result, the supporting organization constitutes an important environmental factor, which affects social capital in community settings. Figure 14 underlines my argument. It shows that organizational social capital provides the environment upon which community social capital develops.

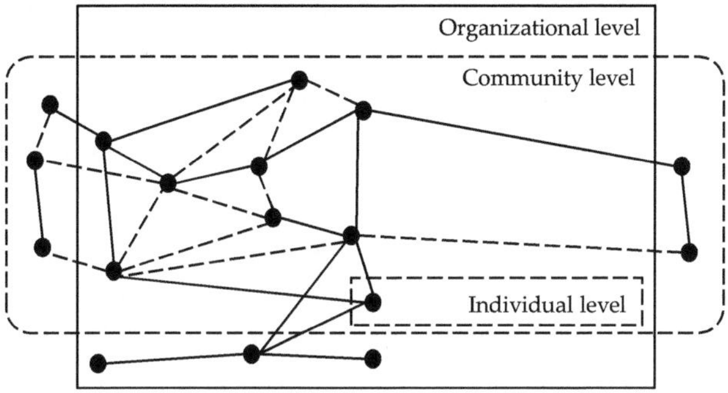

—— Organizational social ties that provide the environment upon which ties in community settings develop
– – Social ties established in the context of corporate innovation community activities

Figure 14: Organizational environment and community social capital

The remainder of this chapter is organized as follows. In presenting the framework depicted in Figure 13, I start by discussing empirical findings as well as theoretical underpinnings for the self-reinforcing nature of social capital. Afterwards, I turn to the multilevel antecedents of innovation community social capital. I start with antecedents at the individual and the organizational level of analysis and close the findings section with a discussion of antecedents of social capital on the community level. I present illustrative quotes[17] of the interviewees as well as related findings of other scholars to underline my arguments.

5.1 The self-reinforcing nature of innovation community social capital

Extant literature offers various arguments for an interrelated and self-reinforcing nature of social capital (e.g., Walker, Kogut, & Shan, 1997). For instance, Carpenter et al. (2012) attest the structural dimension of social capital the character of a path dependent asset (i.e., current network structure influences future network structure). Relatedly, Zaheer and Soda (2009) examine how team composition affects structural

[17] Because interviews have been conducted in German, I translated the illustrative quotes into English. Sometimes, statements had to be slightly adapted to maintain anonymity of the interviewees or for the sake of clarity (e.g., some community members used a notion of innovation community that was specific to the organization that supported the community). None of the changes made affected the meaning of the illustrative quotes.

holes and find that past team cohesion of group members negatively affects structural holes. Often, a basic assumption of scholars who argue for a reinforcing character of social capital is that social capital develops as individuals interact. This facet is mirrored in the following citation of Leana and van Buren: "Organizations wishing to enhance their stores of social capital can do so through employment practices that promote stability among members" (1999: 544). In other words, as stability among organizational members is associated with repeated interaction with other organizational members, it is also related to the development of social capital among members of the organization.

Data also suggests that social capital builds over time as individuals interact. For instance, quotes #1-3 in Table 4 of employees of the toy manufacturing, the automotive as well as the automation company address the relevance of stabile workforces for the development of social capital. The interviewee in quote #4 emphasizes the self-reinforcing nature of social capital as he describes the corporate network as a snowball system. Apart from these four illustrative quotes, I observed this issue during community workshops. For instance, community members who joined the temporary employment company just prior to community workshops had to build trusted relationships with other community members before they have been able to exchange ideas and knowledge efficiently. Collins and Smith (2006) present related findings and argue that commitment-based HR practices (i.e., practices that emphasize mutual long-term relationships between employees and organizations) are positively related to a social climate of trust, cooperation, as well as shared codes and languages.

Table 4: The self-reinforcing nature of social capital

Illustrative quotes: **The self-reinforcing nature of social capital over time**
1. "Fluctuation is not an issue of central importance at our company. Many employees are working for the company for an impressive amount of time. This holds even for the top management of the organization. As a consequence, the social networks between employees of our organization had the opportunity to evolve over a long period of time." *Community manager (Case C: Toy manufacturing)*
2. "We have strong leadership ranks which are closely linked [...]. In many cases, these managers work for our company for many years and developed their networks during this time." *General manager (Case B: Automobile)*
3. "If you work for the same company for many years, you know a lot of people you can ask for help." *Community member (Case D: Automation)*
4. "Our corporate network is a positive snowball system." *Community member (Case D: Automation)*

Based on the five case studies presented in this part, I propose that the self-reinforcing characteristic of social capital is a consequence of five interrelations between the three dimensions of social capital. I refer to Nahapiet and Ghoshal (1998) and distinguish between structural, relational and cognitive social capital. Structural social capital describes the structure of relationships and is regularly measured by interaction patterns in terms of network size or network cohesion (Carpenter et al., 2012). The relational dimension captures trust (Tsai & Ghoshal, 1998) as well as frequency and closeness of interaction (Lechner et al., 2010) and, hence, comprises the quality of social relationships (Moran, 2005). Finally, the cognitive dimension describes shared visions, missions, and languages among community members (McLure Wasko & Faraj, 2005; Nahapiet & Ghoshal, 1998)[18].

In order to explain the self-reinforcing nature of social capital in more detail, I present my findings with regard to five relationships between different dimensions of social capital in the following. First, I suggest that many social ties between community members (i.e., the structural dimension of social capital) provide a basis for intense interaction and are positively related to shared visions and languages (i.e., the cognitive dimension of social capital) within the community. The quotes cited in Table 5 underline my argument. For instance, quote #1 of a community member at the conglomerate suggests that interaction with others is associated with the development of a shared identity, an aspect that is closely related to a shared vision. In other words, people develop a feeling of belongingness while they interact via

[18] For a detailed discussion on the three dimensions, see part II.

social ties. The second quote describes how a shared vision of community members develops as members work and interact during community activities. As another example, the manager of the community supported by the temporary employment company decribes the importance of intense interaction during workshops in order to develop a shared language (quote #3).

Table 5: The effect of structural on cognitive social capital

Illustrative quotes: **Structural social capital → Cognitive social capital**
1. "I am convinced that a major motivation of people to engage in networking activities is to build a shared sense of identification." *Community member (Case E: Conglomerate)*
2. "At the end of the community process, we had established a clear goal that everybody in the community aimed to achieve." *Community member (Case D: Automation)*
3. "Because we used methods that forced us to interact heavily during community workshops (e.g., LEGO Serious Play), we concocted a plan from the very beginning. The whole community used the same terms – even across different groups within the community." *Community manager (Case A: Temporary employment)*

Second, empirical data indicated that numerous interaction ties between individuals are associated with trust within the community. Table 6 offers illustrative quotes to explain the relationship between structural and relational social capital. The first and the second quote originate from community members at the conglomerate and at the automation company. The quotes describe how trust emerges as individuals work together and interact in their community networks. During these processes, individuals get to know each other and gather experience with regard to reliability of their peers. This issue is also reflected in the third quote, which was stated by a community member in the context of a group discussion. In reflecting a past project that was accomplished by the community, the interviewee expressed his trust towards the community that is based upon experiences made during interaction in community settings.

Table 6: The effect of structural on relational social capital

Illustrative quotes: **Structural social capital → Relational social capital**
1. "Of course, trust develops while people interact and work together. [...] If you are motivated to contribute to working processes and willing to learn from others, you can ask them for help. It is all about doing things together!" *Community member (Case E: Conglomerate)*
2. "Trust emerged over time during community work. [...] The more often you collaborate, the more you trust other community members. At the beginning, it is more difficult: Who is that guy? To what department does he belong?" *Community member (Case D: Automation)*
3. "There is an impressive amount of trust among community members. We know that we are all excellent; everybody has contributed and attended our innovation process. If you ask anybody in the community about our work, I believe that everybody here is able to give you a competent answer." *Community member (Case C: Toy manufacturing)*

Third, cognitive social capital is associated with relational social capital. In other words, community members who share visions and languages also build trusted relationships as they have the feeling that they collaborate in order to achieve the same ends. The illustrative quotes #1 and #2 in Table 7 provide empirical evidence for this theoretical line of argument. The interviewee cited describes the importance of a shared language to develop a close relationship to other community members. I furthermore observed the relationship between cognitive and relational social capital in community workshops. While interaction between community members at the beginning of a new community project was often characterized by uncertainty between community members, a more concrete picture of the desired project outcome (i.e., cognitive social capital) in later stages was also associated with closer collaboration and trusted knowledge exchange that was embedded in the shared vision among community members.

Table 7: The effect of cognitive on relational social capital

Illustrative quote: **Cognitive social capital → Relational social capital**
1. "In the process of working together, we started to talk about the same things. This is a crucial process to get to know the other community members better! You understand each other far better and this was very important, at least in my opinion." *Community member (Case A: Temporary employment)*
2. "A shared language has top priority – it enables successful communication [...]. Without a shared language, it is hard to collaborate successfully." *Community member (Case A: Temporary employment)*

Furthermore, cognitive and relational social capital affect structural social capital mirror-invertedly which leads to a self-reinforcing cycle of the three dimensions of social capital. More precisely, I propose that shared visions and languages lead to more intense interaction with other community members (i.e., structural social capital). As the illustrative quote #1 in Table 8 of an employee of the automotive company suggests, people who share a common goal or language exchange knowledge more easily. While the interviewee describes this relationship with regard to a known colleague, observations in workshops and on online platforms suggest that this relationship holds also for collectives. In other words, even if two community members never have been in contact, they interact more efficiently if they share the same visions and languages. In addition, people are more motivated to talk to other community members with whom they share cognitive social capital because they have the feeling that interaction with these people helps them to reach their shared goals. Quote #2 offers a quote of the community manager at the toy manufacturing company. Asked whether he thinks that it will be difficult to motivate people to become community members, the community manager said that he thinks that high identification with the company (i.e., cognitive social capital) will facilitate recruiting of employees. In more detail, the manager argued that shared visions of people are associated with higher motivation to engage in interaction with other community members.

Table 8: The effect of cognitive on structural social capital

Illustrative quotes: **Cognitive social capital → Structural social capital**
1. "If you have a long history of experience with a colleague and you know how he or she thinks, then interaction with this person is easy." *R&D manager (Case B: Automobile)*
2. "I think that we will have no difficulties in motivating people to engage in community activities. Of course, we have to do a good recruiting job. There is a high identification with our company and because of that, I think we will have little difficulties with regard to this aspect." *Community manager (Case C: Toy manufacturing)*

As mentioned above, data furthermore suggests that relational social capital is associated with structural social capital. In other words, close and trusted relationships within the community are associated with interaction with other community members. Table 9 presents two illustrative quotes that underline this aspect. The first statement shows that relational social capital is associated with opportunities to establish further relationships because individuals tend to introduce close ties to other contacts, for instance during lunch time. The second quote emphasizes the relevance of reciprocity in the context of trusted relationships. As Faraj and Johnson (2011) highlight, indirect reciprocity is a common interaction pattern in community settings. That is, community members that receive ideas from other community members tend to help third community members in order to reciproce the help they received from their trusted ties.

Table 9: The effect of relational on structural social capital

Illustrative quotes: **Relational social capital → Structural social capital**
1. "If I ask a friend of mine who is working for the same company if we meet each other for lunch, she often answers: 'Oh, I have got another appointment with some other colleagues – but you are very welcome to join us'. After the meal, five more people are part of my network within the company. Now, I meet these people for lunch even when my friend is not available. That's the way how networks develop over time." *Community member (Case E: Conglomerate)*
2. "If you always ask people for help but never offer your help to others, you will not succeed in developing trusted relationships to other people." *Community member (Case E: Conglomerate)*

A number of studies support my arguments for a self-reinforcing nature of social capital and interrelations between the three dimensions of social capital. Most prominently, Tsai and Ghoshal (1998) propose that the structural dimension of social capital antecedes the relational and the cognitive dimension of social capital and that the relationship between the structural and the relational dimension is partly mediated by the cognitive dimension. A similar argument is put forward in the seminal work of Coleman (1988) who suggests that dense network structures enable the establishment of effective norms as well as trustful ties, which in turn are associated with increased resource exchange.

In addition, related studies support single relationships I propose in the self-reinforcing cycle of social capital in corporate innovation communities. With respect to the relationship between structural and cognitive social capital, Balkundi and Harrison (2006) study the effects of network density on team viability. Viability captures the "willingness [of team members] to stay together as a team" (Balkundi & Harrison, 2006: 52) and is positively affected by network density. As another example, Ibarra, Kilduff, and Tsai (2005) argue that network structure affects shared visions. In their words: "Individuals' positions in friendship networks can bias perceptions of the environment to the relative exclusion of more objective outside views, potentially reinforcing similar views within friendship clusters" (Ibarra et al., 2005: 365). The relationship between structural as well as relational social capital is based on the premise that the development of closer relationships occurs within existing social relationships (Hansen, 1999). For instance, Kostova and Roth (2003) argue that social capital develops as interaction via social ties takes place and Chua et al. (2008) find that network embeddedness is related to trusted ties. As another example, Ferriani et al. (2009) suggest that familiarity to partners in project entrepreneurship is positively related to trust.

The study of Baba and Walsh (2010) offers insights into the facilitating role of cognitive social capital in knowledge exchange and interaction within networks of relationships. In their study, Baba and Walsh (2010) show how dense networks helped drug companies to introduce breakthrough innovations because these dense networks helped to establish shared understandings of the risks of drug candidates. An assumption of these scholars is that the shared understanding allowed more efficient knowledge exchange. As another example, Bechky (2003) suggests that shared languages among different groups in internal communities of practice facilitate interaction. Similarly, Tsai (2000) identifies an organizational units prior network centrality and its trustworthiness as significant antecedents to new linkage creation.

It is important to note that the self-reinforcing cycle of social capital may also result in a doom loop that destroys social capital. While the quotes presented above describe processes where the cycle results in stronger social capital, the quote cited in Table 10 decribes how the doom loop of social capital works and destroys social capital. That is, if an individual judges another individual as not trustworthy, the focal individual will reduce interaction with the untrusted individual and the community as a whole to a significant extent. The reduced amount of interaction, in turn, is associated with reduced opportunities to rebuild trust and a shared vision in the future.

Table 10: The self-reinforcing destruction of social capital

Illustrative quotes: **The doom loop of social capital**
1. "Trust is a very important topic. I do not know if I have already told you this in some way, but I have turned to be cautious with respect to this aspect. I just experienced too many disappointments. Because of that, I disclose little to my colleagues und I do not expect from others to do so. This, of course, results in a way in superficial communication." *Community member (Case E: Conglomerate)*

To sum up, I propose that relationships between different dimensions of social capital can be described as a self-reinforcing cycle. Interaction ties within the community network (structural dimension) provide opportunities to establish a shared vision and language (cognitive dimension) as well as close and trusted ties (relational dimension). Cognitive social capital, in turn, affects the relational dimension of social capital positively. Moreover, cognitive as well as relational social capital are associated with further interaction in the community network. Figure 15 summarizes my findings. In the remainder of this chapter, I present antecedents of social capital at the individual, the organizational as well as the community level. As I will argue below, these antecedents affect structural as well as cognitive social capital.

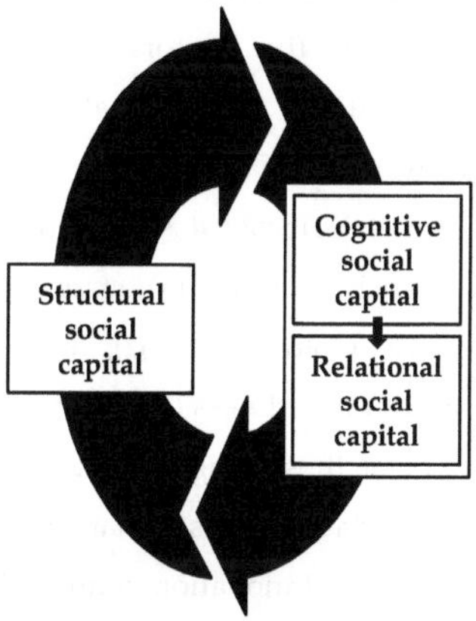

Figure 15: Interrelated dimensions of social capital in corporate innovation communities

5.2 Individual level antecedents of innovation community social capital

My category of individual level antecedents of innovation community social capital captures aspects that are directly related to single community members. Within this category, I identified two sub-categories of antecedents that are of major importance within corporate innovation community settings. First, I propose that personality affects the extent to which an individual establishes social capital in communities and organizations. Second, I argue that motivations of community members to engage in innovation community activities affect network position as well as shared visions and languages of community members. I present empirical as well as theoretical support for both aspects in the following.

5.2.1 Personality traits

As my preliminary analysis revealed that personality affects innovation community social capital, I referred to the literature on personality psychology in order to identify a theoretical basis upon which I am able to analyse the data gathered in a structured way. While the literature offers numerous conceptualizations of personality, the Five Factor Model (FFM; also referred to as the "Big 5") has received most attention (Goldberg, 1992). The FFM, which I use in this study to analyse personality, distinguishes between five factors of personality: (1) *Openness to experience* describes aspects of creativity and originality (Mowen & Spears, 1999); (2) *agreeableness* represents "the warmth of feelings toward others" (Brown, Mowen, Donavan, & Licata, 2002: 112); (3) *emotional stability* refers to the steadiness of a person's emotional mood (Saucier, 1994); (4) *extraversion* captures whether an individual is silent or talkative (Goldberg, 1992); and (5) *conscientiousness* covers aspects related to reliability and thoroughness (Goldberg, 1993)[19].

My data revealed that all of these "Big 5" factors affect corporate innovation community social capital. First, data suggests that openness to experience affects structural social capital in corporate innovation communities. In other words, open community members build more social ties. Table 11 presents illustrative quotes to underline my argument. For example, the first and the second quote highlight that the tendency of open people to engage in creative interactions helps these individuals to establish a large network within the community. Quote #3 underlines these arguments and explains the strength of open community members to be able to cope with distinct ideas in community settings. This ability also allows open individuals to exchange with a large number of peers in community settings. Quote #4 and #5 also stress this aspect. In community contexts, creative and open people benefit from an ability to accommodate ideas of other community members. This ability allows also interaction with a high number of community members.

---

[19] A detailed review of the literature on the FFM is provided in part IV.

Table 11: The role of openness to experience

Illustrative quotes: **Openness to experience → Structural social capital**
1. "In general, I am an open minded person. [...] I have great fun when I exchange innovative ideas with other colleagues." *Community member (Case A: Temporary employment)*
2. "I think I am very open minded. I am open to approach any novel task and try to think about ways to find a solution." *Community member (Case C: Toy manufacturing)*
3. "In community settings, you need to be able to accept opinions of others. This is the most important capability to me." *Community member (Case D: Automation)*
4. "To work efficiently in community settings, you need to be open and to be able to accommodate ideas of others." *Community member (Case E: Conglomerate)*
5. "You need to be open. You need to be able to come up with ideas and to discuss ideas with others." *Community member (Case E: Conglomerate)*

Second, interviewees reported that the personality trait of agreeableness (i.e., likeableness) constitutes a major antecedent of structural social capital. The quotes in Table 12 summarize my findings. Quote #1 and #2 highlight a general tendency of individuals to prefer interaction with agreeable people. As a result, agreeable personalities are able to establish a higher number of social relationships in community settings. The third quote presents a statement of a community member in reaction to the question related to his strengths in community work and underlines the importance of agreeableness to interact successfully with a large number of peers in community settings. The fourth quote discusses the advantage of agreeable people to avoid unnecessary controversy. This is beneficial because dissensions between community members may lead to destruction of social capital. Because of the absence of hierarchical structures (Fleming & Waguespack, 2007), agreeableness is especially important in community settings. As quote #5, #6, and #7 emphasize, respect, trust, and willingness to compromise are important in such environments and enable intense interaction and the creation of many network ties. Quote #8 describes the mediating role of the structural dimension in the relationship between agreeableness and trusted ties. As the interviewee describes, agreeableness is associated with interaction and interaction, in turn, enables trusted relationships.

Table 12: The role of agreeableness

Illustrative quotes: **Agreeableness → Structural social capital**
1. "I can't stand harsh and rude people [...]. Therefore, I don't develop deep social relationships with people who possess these personality traits." *Community member (Case A: Temporary employment)*
2. "I am a very communicative and agreeable person. This aspect has helped me to find a lot of friends and to establish a big network within the company." *Community member (Case C: Toy manufacturing)*
3. "I am a friendly person and, in very most cases, nice to my colleagues." *Community member (Case D: Automation)*
4. "If I can't stand people, I do not show this during conversations. I keep these things for myself. I think I am good at hiding antipathies." *Community member (Case C: Toy manufacturing)*
5. "In community settings, it is important to treat every community member with respect – irrelevant of the hierarchical position." *Community member (Case C: Toy manufacturing)*
6. "I think it is important to be tolerant in community settings. You have to be able to allow others to vocalize their ideas. You need to be able to respond to others." *Community member (Case E: Conglomerate)*
7. "In community settings, you need to be able to compromise." *Community member (Case E: Conglomerate)*
8. "I think that I am likeable and this is beneficial for building trusted relationships. Because of that, many people share information and knowledge with me – these people do not have to do this and I appreciate it very much." *Community member (Case E: Conglomerate)*

Third, I propose that emotional stable individuals establish more ties with other community members than their anxious counterparts. In other words, emotional stability is positively related to structural social capital. Illustrative quote #1 in Table 13 highlights the importance of self-confidence in community settings. The interviewee cited in quote #2 describes that apprentices who took part in community activities often were too silent because they did not have the courage to express their ideas. I observed this aspect between emotional stability and network size of individuals also at community workshops, where self-confident individuals succeeded in bringing up new ideas and discussing their arguments with a large number of peers, while unsettled community members had more problems to contribute to community work. The ability of emotional stable individuals to establish social relationships traces back to two distinct lines of arguments. First, these personalities experience fewer barriers to articulate their ideas to others because they trust in their abilities to contribute to innovation processes. Second, I argue that because emotional stable people are often more satisfied, other community members enjoy interaction with these individuals and, as a consequence, are more likely to approach these people.

Table 13: The role of emotional stability

Illustrative quotes: **Emotional stability → Structural social capital**
1. "It is important that you feel certain about what you are doing [...]. It is important that you don't allow others to let it get to yourself." *Community member (Case C: Toy manufacturing)*
2. "In my opinion, the apprentices did not engage enough in interaction in community settings." *Community member (Case D: Automation)*

Fourth, analysis of interviews and observations of online and offline collaboration in community settings revealed that extraversion is an important antecedent of structural social capital. In essence, the personality trait of extraversion determines whether an individual is at ease in getting in contact with unknown people (Goldberg, 1992). As a result, extraverted individuals often engage more in networking activities than others (Forret & Dougherty, 2001). My argument with regard to the relationship between extraversion and structural social capital is illustrated in the quotes in Table 14. Quotes #1-#4 show the capability of extraverts to establish relationships with unknown community members. As a result, I suggest community members who possess this personality trait form more social relationships in community settings. Quotes #5 and #6 highlight the relevance of communication skills in establishing social relationships in community settings. Extraversion and intense communication are closely related. As such, these quotes underpin my argument for a relationship between extraversion and network size of community members. Quote #7 in Table 14 highlights an additional aspect. Extraverts are especially suitable to innovation community contexts because these people are able to mask out hierarchical structures while interacting with other community members. As community members belong to different hierarchical levels, the ability to establish relationships with other community members who belong to different hierarchical levels positively affects the ability to build a huge network within the community.

Table 14: The role of extraversion

Illustrative quotes: **Extraversion → Structural social capital**
1. "People differ with regard to their ability to get in touch with new people. Some are active from the very beginning, and others need time until they participate actively." *Community member (Case A: Temporary employment)*
2. "I love to meet unknown people. For sure, you need a few moments until you recon up the character of somebody. But I don't have a problem with that." *Community member (Case A: Temporary employment)*
3. "I think I am very good at dealing with people." *Community member (Case C: Toy manufacturing)*
4. "I approach unknown people very regularly." *Community member (Case D: Automation)*
5. "There are communicative people and there are people who are not very communicative. I belong to the latter group. It is not easy for me to establish social relationships." *Knowledge & idea manager (Case B: Automobile)*
6. "I have also worked in the service sector for a long time. Related to that, I am good at communicating. I like working with and for people." *Community member (Case C: Toy manufacturing)*
7. "I would describe myself as uncomplicated [...]. And I live my life according to this method: I get in contact with people in an open way [...]. I am an anarchist who does not care much about hierarchy. I talk to ordinary people and the top management in the same way." *General manager (Case B: Automobile)*

Fifth, my data suggests that conscientiousness affects structural as well as cognitive social capital positively. The four quotes in Table 15 (#1-#4) show the achievement oriented nature of conscientious community members, a facet that is also emphasized in the personality psychology literature (Judge, Higgins, Thoresen, & Barrick, 1999). With regard to innovation community social capital, I propose based on empirical observations that conscientiousness is beneficial for structural as well as cognitive social capital. First, conscientious individuals interact with many other community members because they are highly motivated to perform extraordinary and to integrate ideas of a large number of peers in order to compile the best possible output. Second, because of their ambitious nature, conscientious community members identify heavily with the community task and perceive it as a vision, which they share with the community.

Table 15: The role of conscientiousness

Illustrative quotes: **Conscientiousness → Structural & cognitive social capital**
1. "I think community members are people who are not satisfied with cookie cutter approaches. These individuals are looking for challenges: 'What can I do better? What can I do different in order to get better?" *Community manager (Case C: Toy manufacturing)*
2. "I am a perfectionist – and I want to be that way both in business as well as private contexts. I can't stop trying to do everything as good as possible. I am constantly putting my foot down. I want to achieve 100% - or at least something close to it." *Community member (Case E: Conglomerate)*
3. "I'm missing some kind of carelessness that allows me to accomplish tasks not perfectly." *Community member (Case D: Automation)*
4. "I am an extremely ambitious person [...]. If I want to achieve something, I always go for the 100% [...]. And in most cases, I reach my goals." *Community member (Case E: Conglomerate)*

Several extant researchers have already emphasized the importance of personality as an antecedent of social capital. For instance, Klein et al. (2004) find that emotional stability affects centrality of individuals in advice networks within teams. Similarly, Oh and Kilduff (2008) find that personality influences whether an individual owns a network position associated with brokerage or not. Likewise, Thompson (2005) finds that proactive personalities are superior in building networks and, as a consequence, perform better in their jobs. Recently, Sasovova et al. called for more research in this context and argued that "network research, at least since the 1970s, has largely eschewed analytical approaches that direct attention to individual differences, but it is likely to be precisely such individual differences that influence network dynamics" (2010: 640). In response, they study the extent to which the personality trait of self-monitoring (i.e., the degree to which an individual is able to control himself to create suitable self-presentations) affects network positions. They find that self-monitoring is positively related to the number of new friends over time (especially with people from other functional groups), the number of structural holes in the network as well as the maintainance of these structural holes.

Furthermore, my findings with regard to the trait of openness to experience are closely related to the open innovation literature, which often highlights the central role of lead users (i.e., most innovative and creative users) within communities (Mahr & Lievens, 2012). I suggest that many lead users possess the personality trait of openness to experience. Jeppesen and Laursen (2009) argue that lead users exchange ideas in communities more intensely than others do because they hold the relevant knowledge to do so.

5.2.2 Intrinsic and extrinsic motivations

Apart from aspects of personality, data indicates individual motivation to engage in community work as an antecedent of social capital. To structure my analysis in this context, I refer to Janzik (2010) and distinguish between intrinsic and extrinsic motivations of community members. While intrinsic motivation describes individual behavior which is motivated by an activity itself, extrinsic motivation describes activities based on external rewards (Deci, 1972).

I argue that intrinsic motivation is positively related to structural as well as cognitive social capital. The quotes in Table 16 summarize my empirical analysis. With regard to a relationship between intrinsic motivation and network size, the first quote highlights that intrinsic motivation may lead to interaction with other community members even if personality (e.g., introversion) works against approaching unknown individuals. Quote #2 highlights the desire of other community members to collaborate with intrinsically motivated counterparts. Hence, I suggest that intrinsically motivated community members are able to form more social ties in community settings. In quotes #3, #4, and #5, intrinsic motivated interviewees highlight the importance of fun during community activities. Related to these quotes, I observed during community activities that cheerful people with good humor succeeded more easily to interact with previously unknown individuals. Because having fun is an important intrinsic motivation to engage voluntarily in community activtities (Janzik, 2010), community members seek interaction with others who are also in for pleasure. Furthermore, there is empirical evidence that intrinsically motivated individuals establish cognitive social capital with other community members more easily. In essence, intrinsic motivation is related to interest in the community process itself. I observed at community workshops that intrinsically motivated individuals had a strong interest in the outcome of community processes and felt closely attached to community work. Thus, I propose that intrinsic motivation is positively related to shared visions (i.e., cognitive social capital) in community settings.

Table 16: The role of intrinsic motivation

Illustrative quotes: **Intrinsic motivation → Structural & cognitive social capital**
1. "I love to live the calm life. I don't like too much action and to meet too many people. I dislike crowded places [...]. However, if I am interested in a certain topic, I even go to these places." *Community member (Case C: Toy manufacturing)*
2. "Community members should be interested in topics the community works on [...]." *Community member (Case C: Toy manufacturing)*
3. "When I engage in community activities, it is important for me to have fun." *Community member (Case E: Conglomerate)*
4. "A community workshop is just more fun than sitting the whole day in the office." *Community member (Case E: Conglomerate)*
5. "The nice atmosphere, the stimulating collaboration in the community is funny!" *Community member (Case D: Automation)*

In addition, data suggests that extrinsic motivation affects structural social capital (i.e., network size of community members). Quote #1 (Table 17) describes a general desire of an interviewee to benefit from participation and interaction within the community. Other quotes in Table 17 highlight the relevance of motivations related to reputation (quote #2) and prestige (quote #3) in community settings. The interviewees cited in quotes #4 and #5 address their motivations to broaden networks within the community explicitly. I propose that extrinsic motivation is positively related to structural community social capital. For instance, I suggest that a desire to benefit from reputation and prestige prompts community members to interact with many other community members. Obviously, those who join the innovation community because they aim to build social relationships will also interact with previously unknown peers if opportunities arise.

Table 17: The role of extrinsic motivation

Illustrative quotes: **Extrinsic motivation → Structural social capital**
1. "I don't have concrete expectations towards the company related to my participation in community activities. However, I have a small hope that it will be advantageous in the future." *Community member (Case C: Toy manufacturing)*
2. "I want to establish a reputation within the company to be responsible for issues related to the topics we deal with in the community. I think my colleagues even expect me to know something about these issues and to be able to distribute information in this context." *Community member (Case C: Toy manufacturing)*
3. "I don't know how to describe it... let's just call it prestige. There are moments when it is beneficial if someone mentions your name. Because I participate within the community, I can exhibit experience and engagement in this context." *Community member (Case E: Conglomerate)*
4. "I hope that the community work helps me to get in contact with people and to find new friends. I want to enlarge my network in the community." *Community member (Case C: Toy manufacturing)*
5. "The ability to establish a network within the company in the course of community work was a major motivation to participate. The knowledge of experienced colleagues just helps a lot." *Community member (Case D: Automation)*

My findings with regard to extrinsic and intrinsic motivations of community members as antecedents of social capital in corporate innovation communities are closely related to the efforts of many others who have analyzed voluntary participation of individuals in innovation community activities (Lerner & Tirole, 2002; Muhdi & Boutellier, 2011). As I have already mentioned, intrinsic as well as extrinsic motivations have been identified as relevant constructs in community settings. For instance, there is evidence for intrinsic facets such as fun and enjoyment (Nov, 2007) as well as extrinsic facets such as reputation (Roberts, Hann, & Slaughter, 2006) as motivational aspects in community settings. I suggest that a social capital perspective offers an interesting contribution to the arguments of these scholars as it allows a more nuanced examination of the consequences of intrinsic and extrinsic motivations in community settings.

To sum up, data indicates that, on the individual level, personality traits of openness to experience, agreeableness, emotional stability, introversion and conscientiousness as well as intrinsic and extrinsic motivations of community members antecede structural social capital. In addition, the personality trait of conscientiousness and intrinsic motivation affect the cognitive dimension of social capital. Figure 16 summarizes individual level findings of the framework presented.

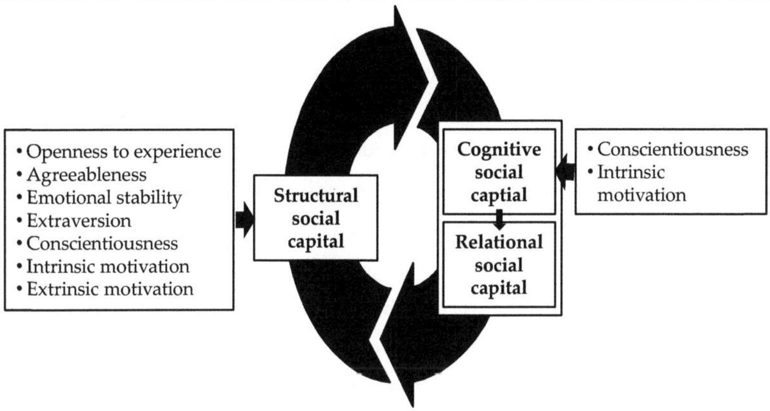

Figure 16: Individual level antecedents of social capital

5.3 Organizational level antecedents of innovation community social capital

Organizational level antecedents of social capital capture those antecedents which affect supporting organizations as a whole and as such, determine the environment in which community social capital develops. This category captures antecedents that affect all members of the organization that supports community activities. Note that this category is of specific importance in communities that consist of a larger group of members of the supporting organization[20]. However, organizational level antecedents are also relevant in innovation communities consisting of members from different organizations. Here, each participating company influences community social capital via antecedents situated at its own organizational level. I identify two major sub-categories (i.e., corporate culture as well as formal and informal organization) on this level and discuss my findings in the following.

[20] It is important to emphasize that the communities within my sample all meet this criteria.

5.3.1 Corporate culture

My data reveals that a central antecedent of social capital in corporate innovation communities at the organizational level is corporate culture. Leana and van Buren argue that an organizational culture provides an "overarching philosophy and corresponding norms within which different individuals enact that philosophy" (1999: 545). My data provides evidence that two facets of corporate culture affect social capital in corporate innovation communities. First, I argue that an organizational culture that supports collaboration between employees affects the structural (i.e., number of ties) as well as the cognitive dimension (i.e., shared visions and languages) of social capital. Quotes #1-#3 in Table 18 highlight these aspects and emphasize the role of collaborative cultures in facilitating the establishment of network ties as well as in developing a shared vision among employees and community members. Quote #4 pulls in a similar direction and highlights that the collaborative corporate culture of the conglomerate is both associated with feelings of belongingness as well as with interaction via network ties within the company as organizational members help each other regularly. Employees of the supporting organization transfer these behavioral patterns to community settings, and as such, collaborative cultures affect network structures within corporate innovation communities.

Based on my data, I identified two aspects that affect the development of a collaborative corporate culture. First, quote #5-#7 highlight that identification with the products of the company supports such a culture. Quote #5 and #6 decribe the benefits of emotional products such as toys or cars in this context. As the interviewees signify, emotional products create a feeling of belongingness to the company as well as the workforce. The seventh quote of an employee of the automation company argues in a similar direction and shows how interest in technological products can create such a culture as well. Second, quotes #8-#10 emphasize that organizational growth is detrimental to collaborative cultures. The negative impact of organizational growth on collaborative cultures is related to the arguments of the role of workforce stability in establishing corporate social capital I have highlighted before (Leana & van Buren, 1999).

In a nutshell, I argue based on empirical evidence that a collaborative culture is beneficial for structural social capital because such a culture legitimizes interaction with previously unknown colleagues. Furthermore, a collaborative climate is positively related to cognitive social capital because it leads to shared identification, visions, and languages among employees and community members.

Table 18: The role of collaborative corporate culture

Illustrative quotes: **Collaborative corporate culture → Structural & cognitive social capital**

1. "At our company, we all pull together." *Employee (Case A: Temporary employment)*
2. "I was never a lone fighter in the company. We do everything in teamwork." *Community member (Case D: Automation)*
3. "One reason why I am working for this company is because I enjoy the climate and the work here. There is not an extreme competition between employees – we do not try to score off one another." *Community member (Case C: Toy manufacturing)*
4. "I think that being an employee of this company is related to some form of trust. […]. It's somehow like the famous saying goes: 'I scratch your back, and you scratch my back.' And I think that's ok." *Community member (Case E: Conglomerate)*
5. "The intensive identification of our employees with the company is beneficial for innovation processes. The identification with and the engagement for the company is incredible. Of course, this is related to the kind of products we sell." *Community manager (Case C: Toy manufacturing)*
6. "This question is related to corporate culture – we have an impressive identification with our products and the company. It is like a spirit among employees: I belong to this company. It is typical for such an emotional product like a car […]. I hope you can feel the enthusiasm." *General manager (Case B: Automobile)*
7. "I very much like the products that our company sells!" *Community member (Case D: Automation)*
8. "I think people reduce their identification with the corporation as an organization grows rapidly." *Community member (Case C: Toy manufacturing)*
9. "I doubt that employees of the company all share common goals, because the organization is far too big. And of course, everybody tries to succeed at his own task." *Community member (Case E: Conglomerate)*
10. "Because the company now has an image to be successful, other characters start to apply for jobs. Sometimes, these people are too much focused on their own careers – and do not care so much about shared goals of the workforce as a whole." *General manager (Case B: Automobile)*

I identified an additional aspect of corporate culture that antecedes social capital within corporate innovation communities. More specifically, I suggest based on empirical findings that a corporate culture that emphasizes the importance of innovation affects structural as well as cognitive social capital of corporate innovation communities. Employees working in such corporate climates are able to freely assign a certain amount of working time to promising, innovative issues. The first and second quotes in Table 19 highlight the importance of an innovative culture for social capital in general. The interviewees cited point to the freedom that is associated with such cultures. Quotes #3-#5 in Table 19 highlight the role of time pressure in innovation community settings. Employees are unable to build social relationships and shared visions in the community if their daily duties do not allow them to join the other community members and their activities. The sixth quote highlights the drawbacks of limited time in innovation processes in general. As the interviewee employed by the automotive company describes, overcrowded schedules are destructive in the context of innovative processes in organizations because these schedules reduce the opportunity to interact with others and to build networks. Quote #7 of a community member at the temporary employment company unveils an additional aspect. A corporate culture that focuses too much on efficiency as opposed to innovation is related to a tendency to reduce face-to-face interaction and to substitute it with interaction via e-mail. As I will propose based on my observations later on, interaction via e-mail is not able to generate social capital to the same extent than face-to-face interaction. Finally, the eigth quote describes the importance of boundary crossing innovation processes for structural social capital. In innovative cultures, large cohorts of employees are involved in innovation processes, which provides them the opportunity to establish network ties across departments and hierarchies. In addition, an innovative culture that prompts large cohorts of employees to engage in innovation processes implies that employees perceive innovation as something important. Hence, innovative corporate cultures affect shared visions with respect to community work positively.

To sum up, data indicated that an innovation-oriented culture is associated with freedom of employees to exchange ideas and knowledge in order to participate in community settings. As a result, employees in such companies have more time to engage in community activities and develop more ties within these settings. In addition, employees who work in such settings are more likely to develop a shared vision with other community members because an innovative culture promotes the extent to which employees perceive community outputs as important.

Table 19: The role of innovative corporate culture

Illustrative quotes: **Innovative corporate culture → Structural & cognitive social capital**

1. "Our company is very innovative – the management deliberately provides many play spaces for us." *Community member (Case D: Automation)*
2. "In most cases, we try to work democratically [...]. That's the way how we solve problems." *Community member (Case D: Automation)*
3. "It is important to allow employees to participate at the innovation community – they have to be able to reorganize their work or to buy themselves some time." *Community member (Case C: Toy manufacturing)*
4. "The time pressure is really important in this context. A community workshop takes a whole day – of course you need a principal who supports you in order to engage in such activities." *Community member (Case E: Conglomerate)*
5. "An important barrier is the time. I already missed one community workshop because of that. This is the case because I had other appointments which I am not allowed to and I don't want to miss." *Community member (Case E: Conglomerate)*
6. "It is not because I want to annoy my colleagues – it's just because I have no choice. I just have not enough time – and this is also not good with respect to innovation." *General manager (Case B: Automobile)*
7. "In most cases, we interact via e-mail or face-to-face meetings. Often, I use e-mail because I do not have enough time." *Community member (Case A: Temporary employment)*
8. "At our company, innovation processes cross the boundaries of departments. Sales, marketing, research – everybody is involved." *Community member (Case D: Automation)*

The literature offers several studies that support my arguments with respect to the relevance of organizational culture as an antecedent of social capital. For instance, Inkpen and Tsang (2005) compare intraorganizational and interorganizational social networks and argue that social capital is easier established in intraorganizational networks. According to the arguments of Inkpen and Tsang (2005), a shared corporate culture is positively related to the development of shared visions. As another example, Leana and van Buren (1999) propose that the corporate culture of General Electrics is beneficial for the development of structural as well as cognitive social capital. In their words, "General Electric has developed an organizational culture in which shared commitment is built not necessarily through stable individual relationships but, rather, through an overarching organizational philosophy that emphasizes teamwork" (Leana & van Buren, 1999: 545). Relatedly, Kim, Lee, Zmund, and Bock (2005) find that an appropriate organizational climate significantly affects knowledge sharing in organizations. Brockman, Rawlston, Jones, and Halstead (2010) study antecedents of interpersonal cohesiveness in new product development teams. Interpersonal cohesiveness captures a tendency of attraction to groups and is related to cognitive social capital. Brockman et al. (2010) find that an organizational culture that emphasizes teamwork, cooperation and loyality as well as cross-functional integration is positively related to interpersonal cohesion among team members.

5.3.2 Formal and informal organization

I propose that formal as well as informal organization affect social capital in innovation communities that receive corporate support as well. My findings in this category are related to Gulati and Puranam who observe a reorganization at Cisco and distinguish between formal organization (i.e., "the normative social system designed by managers" (2009: 422)) and informal organization (i.e., "the emergent pattern of social interactions within organizations" (2009: 422)). My interviews and observations stress the role of formal and informal organizations as antecedents of social capital. More precisely, there is evidence that both forms of organizations represent architectures upon those structural social capital in organizations and communities is able to develop.

Table 20 summarizes empirical evidence of the relevance of formal organization as an antecedent of structural social capital. In other words, formal organization determines the formal ties of an employee and, hence, affects network structures to a significant extent. The first and the second quote describe how workflow determines with whom an employee regularly interacts during working hours. Quote #3 stated by an employee of the automotive company emphasizes the crucial role of formal processes that cross boundaries in organizations to establish ties in other units. Furthermore, as the fourth and fifth quotes highlight, project organization works against the existence of structural holes because of the same line of argument. The community member cited in quote #6 descibes the relevance of formal hierarchy for social capital. An employee who never changes the department has fewer opportunities to establish structural social capital as this employee has fewer formal ties within the organization. I also observed the role of formal organization with respect to community social capital at community workshops. At the beginning of a new community project, community members tended to socialize with peers who they knew because of formal hierarchy, projects or processes. Over the course of the project, the tendency to talk to those people who worked for the same department regularly diminished as community members had established social realtionships with other community members as well. Notwithstanding the importance of formal organization in the creation of social capital, quote #7 and #8 emphasize the importance of informal organization that also affects social capital in organizations and innovation communities. I discuss the role of informal communication as an antecedent of social capital in the next paragraph.

Table 20: The role of formal organization

Illustrative quotes: **Formal organization → Structural social capital**

1. "I work together with many people. Sometimes I talk to apprentices and at other times, I have a meeting with the CEO. This allows me to put ideas that are important to me in place." *Community manager (Case C: Toy manufacturing)*
2. "Because of the workflow, I do not get in touch very often with the headquarter. I have more contacts in other branches." *Employee (Case A: Temporary employment)*
3. "We have a process organization that crosses boundaries of departments [...]. About 5,000 employees from various departments work together in these processes. In my mind, it is well organized." *General manager (Case B: Automobile)*
4. "In addition, we solve many problems in project teams. In other words, if an issue has to be solved with employees from different departments, we initiate a project [...]. These projects differ from processes that cross the boundaries of departments: The project teams split apart after they have solved their task." *General manager (Case B: Automobile)*
5. "I cooperate with different departments during projects within the company. Of course, my daily routines affect my network." *Community member (Case D: Automation)*
6. "A colleague of mine works for the same department since he joined the company – of course, his network is limited to some extent as a result." *Community member (Case E: Conglomerate)*
7. "We are not a very hierarchical company. In consequence, employees have a high degree of freedom to collaborate with others and to build informal networks." *General manager (Case B: Automobile)*
8. "I think it is important to have a hierarchical structure that represents a fundamental order in some way. In addition, it is important to have an informal organization as a complementary organization. Without the informal organization, the formal organization would not be stable enough." *General manager (Case B: Automobile)*

As does the formal organization, the informal organization also offers a structure upon which social relationships in community settings are able to emerge. The extent to which informal organizations exist and are able to provide structural social capital is closely related with collaborative and innovative corporate cultures. I provide empirical evidence for informal organization as an antecedent of community social capital in Table 21. The first quote highlights the interplay between formal and informal organizations and stresses that informal organizations affect the number of ties within organizations. In quote #2 and #3, interviewees discuss the relevance of informal networks of employees who spend their leasure time together. The quotes reflect the relevance of these relationships for business issues. The fourth quote summarizes the role of informal settings in the establishment of structural social capital. As stated, informal settings, such as coffee breaks, provide an important opportunity to get in contact with unknown peers. Quote #5 points to the importance of appropriate climates to develop informal ties within the organization.

To sum up, I propose that formal as well as informal organization affects interaction patterns, number of ties, and hence, structural social capital in organizations and communities.

Table 21: The role of informal organization

Illustrative quotes: **Informal organization → Structural social capital**
1. "There is always an official way, and there are the inofficial networks. In my mind, the second aspect is really important at our company." *Community manager (Case D: Automation)*
2. "There are many opportunities to communicate informally: Joined breakfast or lunch, to go for a walk downtown, [...]. There are even groups of employees who go to the bowling center within their leasure time. [...]. Others meet in a bar to drink a beer together every two months. Even if you try not to talk about business issues – you always end up talking about the company." *Community member (Case C: Toy manufacturing)*
3. "I am also in contact with some colleagues in my leisure time." *Community member (Case A: Temporary employment)*
4. "People spend their lunch time together or meet for a coffee – such activities are important to develop your network." *Community member (Case E: Conglomerate)*
5. "I think our organization tolerates communication that skips hierarchical levels. I could speak to my supervisor's boss without being in trouble." *General manager (Case B: Automobile)*

Interviewees further mentioned several corporate activities that support formal as well as informal interaction ties within organizations. While Table 20 and Table 21 describe the role of formal and informal interaction patterns for the development of social capital in organizations and corporate innovation communities, I discuss activities of companies to support formal and informal interaction ties in the following. The arguments summarized in these categories affect formal and informal structures and, hence, are relevant with respect to social capital, too. I categorize my findings similar to Cherns (1976) in social as well as technical activities that support formal and informal ties.

Table 22 summarizes empirical findings with respect to social supporting activities for formal (quote #1-#4) as well as informal network ties (quote #5-#9). First, with respect to support of formal interaction ties, quote #1 and #2 call attention to the role of education as well as training courses, and quote #3 and #4 discuss events that aim to foster formal interaction between employees. Second, with respect to informal interaction, quote #5 highlights the importance of informal pre-seminar activities to support exchange and the creation of ties between participants. Likewise, in the sixth quote, an interviewee talks about informal events organized by a company and describes the relevance with respect to network structures within the organization. Quotes #7-#9 underline the role of sports events as well as coffee places to support informal interaction ties. To sum up, I suggest based on empirical evidence that social supporting activities affect formal and informal interaction ties in organizations and corporate innovation communities.

Table 22: The role of social supporting activities of organizations

Illustrative quotes: **Social supporting activities of formal and informal interaction ties**
1. "The company organizes an education program for top managers. Here, 20 people learn together for 18 months. This education program establishes the strongest networks in our company – these 20 people never loose sight of themselves." *General manager (Case B: Automobile)*
2. "I regularly meet colleagues from different branches across Germany at training courses [...]. The company seminar is an additional opportunity to meet these people again." *Employee (Case A: Temporary employment)*
3. "There are roadshows where the top management or ordinary employees are invited. During these events, certain departments present themselves and their work. I would describe these events as formal."*General manager (Case B: Automobile)*
4. "There are meetings where different departments participate. Of course, such events affect network structures." *Community member (Case D: Automation)*
5. "At some seminars, you meet the evening before the event starts. You go to lunch, to a bar or to the bowling center [...]. In informal settings, people chat and build trust." *Community member (Case E: Conglomerate)*
6. "Some events at the company are 100% fun within great environments. Here, employees enjoy a loose atmosphere and get in contact with others. In consequence, employees talk to others they have never seen before – in some cases they have never been in contact with their peers, in others the contact was limited to e-mail or phone." *Knowledge & idea manager (Case B: Automobile)*
7. "There are sports tournaments organized by the company, for instance soccer or volleyball. Another example is the 24 hour run in the company." *Knowledge & idea manager (Case B: Automobile)*
8. "The company offers a lot for its employees: Sauna, sports groups, soccer and table tennis tournaments [...]. In this context, the engagement of the company is impressive." *Community member (Case D: Automation)*
9. "In the R&D department, we have small coffee places to provide researchers and developers a room where they can meet each other in a relaxed atmosphere and work on specific topics. Later on, when the topic is more specific, a formal organization takes over." *R&D manager (Case B: Automobile)*

Table 23 shows my empirical findings with regard to technical supporting activities of formal and informal interaction ties. In this context, quote #1-#4 offer examples for technical support of formal communication. The exemplary citations describe the value of electronic directories (quote#1) and intranets (quote #2-#3) to find information with regard to formal communication practices within companies. Furthermore, technical support of informal communication is positively related to the number of ties within the organization. For instance, quote #4 of a community member at the conglomerate highlights the role of intranets in establishing informal relationships, for instance if employees are able to build virtual groups that do not represent formal structures of the organization. The fifth quote also describes the role of technology in supporting informal relationships, for instance if employees are able to make lunchtime appointments on the intranet. Interestingly, as represented in quote #6, my data suggests that technical support is a complement in contrast to a substitute of social support. As highlighted by the interviewee, face-to-face communication has a number of advantages in establishing relationships and social capital. In a nutshell, I propose that technical support affects formal and informal networks, which build the basis for community social capital to develop.

Table 23: The role of technical supporting activities of organizations

Illustrative quotes: **Technical supporting activities of formal and informal interaction ties**
1. "The company provides an electronic directory of all employees. For example, you just fill in a name and the software tells you the site and the department where the focal person works." *Community member (Case E: Conglomerate)*
2. "You can use the intranet to search for specific terms. In most cases, the intranet offers you a contact person for your problem." *Community manager (Case D: Automation)*
3. "Our intranet provides an impressive amount of information: Process instructions, implementing provisions, [...]. In addition, the intranet informs the employees about the activities of our CEO." *Employee (Case A: Temporary employment)*
4. "Employees have the opportunity to build virtual groups in the intranet of our company. For instance, there is one group for students who work for the company beside their studies." *Community member (Case E: Conglomerate)*
5. "The company provides a platform that allows employees to arrange appointments for lunch. It's called 'knowledge at lunchtime'. You can register and provide some basic information about your capabilities and interests and get in contact with colleagues. I think the platform has many advantages because you would never approach an unknown colleague offline." *Community member (Case D: Automation)*
6. "I think it is difficult to establish a network exclusively on the basis of online tools. You can of course foster existing networks – and this is really helpful." *Community member (Case E: Conglomerate)*

Several scholars have already emphasized the relevance of formal as well as informal organization as an antecedent of social capital. For instance, Brass and colleagues argue that formal processes and structures "locate individuals and groups in physical space and in particular points in the organization", whereby "restricting their opportunity to interact with some others and facilitating the interaction with still others" (2004: 796). Likewise, Inkpen and Tsang (2005) argue for the strength of formal hierarchy and processes to establish connectivity between two members of a focal organization. Leana and van Buren (1999) propose that institutionalized processes with specified roles can foster social capital as these roles make individuals interchangeable to a certain extent. Brockman et al. (2010) highlight the importance of informal organizations and find that excessive formalization of interaction within teams is negatively associated with social cohesion because it inhibits interaction between team members. Also related to my findings, Gulati and Puranam (2009) demonstrate the interplay between formal and informal organization as they show that after a reorganization at Cisco, the former formal organization remains in the company as an informal organization.

In a nutshell, empirical evidence suggests that a collaborative and innovative corporate culture as well as formal and informal organization antecede corporate innovation community social capital. In the framework offered in this chapter, these facets reside at the organizational level of antecedents of community social capital. Figure 17 summarizes the findings. I turn to my findings with respect to antecedents of social capital on the community level in the following.

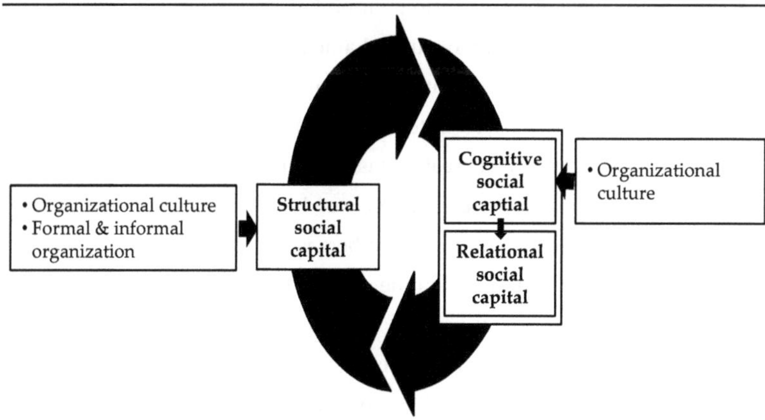

Figure 17: Organizational level antecedents of social capital

5.4 Community level antecedents of innovation community social capital

My category of community level antecedents captures antecedents that affect innovation community members as a collective. I divide my findings into three categories: (1) Creative task design, (2) community leadership, and (3) workshop and platform design.

5.4.1 Creative task design

Task design is an important aspect in innovation community settings (Müller-Seitz & Reger, 2009; von Hippel, 1994). In this context, scholars regularly argue for the advantages of creative tasks because the innovative character of community work constitutes an important motivation of community members (Lakhani & Wolf, 2005). My data suggests that creative task design in community settings is a double-edged sword. I propose that creative task design affects cognitive community social capital both positively as well as negatively. Table 24 presents empirical support for both relationships. With respect to positive effects of creative tasks on cognitive community social capital, quote #1 and #2 describe how creative task designs motivate members to engage voluntarily in interaction in community activities. Analysed through a lens of social capital theory, the creative task allowed the members cited to establish a shared vision with other community members because they heavily identified with the innovative characteristics of community activities. This shared vision, in turn, is associated with interaction with other community members in order to realize desired outcomes. Quote #3 also provides evidence in this direction and describes how a creative task design positively affects interaction between community members. As the interviewee highlights, creative tasks inhibit unfavorable arguing between community members who have different opinions how to approach a clearly defined output. In other words, creative tasks are beneficial for cognitive community social capital because community members are able to shape their vision of community work by themselves.

Yet, as quote #4-#7 highlight, creative tasks also have drawbacks with regard to cognitive social capital. By definition, creative processes are characterized by a lack of an a priori defined output. Related to that, innovation community members may have difficulties in developing a shared vision, especially in the early stages of innovation community processes. I also observed the negative influence of creative task design on cognitive community social capital during workshops. Often, community members discussed heavily how to interpret the task the company asked them to solve in the first workshops. As the community moved forward and collected and evaluated ideas, members also succeeded in generating a shared vision on what they aim to achieve.

Because creative task design has both benefits and drawbacks, I suggest that the relationship between creative task design and shared visions within the community follows an inverted U-shaped pattern[21]. Very uncreative tasks are not suitable to innovation communities because members are not able to identify with that task. At the same time, too open and creative tasks trigger intense discussions among community members at the beginning of a community project because its very hard for community members to figure out which kind of output they are asked to develop.

[21] My qualitative evidence with regard to the U-shaped relationship between creative task design and community social capital is not further investigated in this dissertation.

Table 24: The role of creative task design

Illustrative quotes: **Creative task design → Cognitive social capital**

1. "I really liked the announcement of the new initiative of the innovation community. The substance of community work was described in a very open and general way. I liked that!" *Community member (Case C: Toy manufacturing)*
2. "I would describe the work in the community as exciting – you never know where your journey goes along and how the outcome will look like. And it was fun to solve that task within the community – it was good company." *Community member (Case C: Toy manufacturing)*
3. "I think it is beneficial that we do not have a clearly defined output we are going to achieve. If there is a clearly defined outcome, you end up with discussions of community members who have different opinions. The more precisely the desired output is defined, the more intense are the frictions between community members." *Community member (Case E: Conglomerate)*
4. "Community work is challenging, fascinating – and unsuspecting. At the beginning, I didn't have a clue what we are going to achieve." *Community member (Case E: Conglomerate)*
5. "I expected the community work to be different. I thought we would elaborate on our task in more depth, more like a classical project. I still don't understand how we are going to mix up all the ideas we have now collected." *Community member (Case E: Conglomerate)*
6. "We had intense discussion on our mission at the beginning of the last community project. I think this was the case because it is difficult for some people to identify a goal if it is not clearly defined." *Community member (Case D: Automation)*
7. "It was difficult for me to understand our goal at the beginning. But I got it at the second workshop." *Community member (Case D: Automation)*

5.4.2 Community leadership

Based on empirical observations, I propose that appropriate leadership within innovation communities affects cognitive social capital positively. While communities operate outside hierarchical structures, some members within the community take the position of a leader. As Fleming and Waguespack put it: "Despite their bazaarlike, egalitarian, argumentative, unplanned, chaotic appearance, open innovation communities rely heavily on strong leadership to function effectively and to resist splintering, forking, and balkanization" (2007: 165).

I suggest based on empirical data that these leaders take an important role in establishing cognitive community social capital. Related to my arguments with respect to creative task design, I propose that community leaders play a pivotal role in supporting community members to identify a shared vision of their community work. The quotes cited in Table 25 underline my arguments. The first, second and third quote point to the importance of community leaders in general as well as the role of these leaders in navigating community work towards a desired outcome. In other words, community leaders support community members to identify their shared vision of the creative task they solve in a joint effort. In this context, an important task associated with community leadership is the structuring and target orientation of community activities. For instance, the fourth and fifth quote describe how a short summary of intermediate results and preparation of workshops helped community members to structure their ideas and to establish shared thoughts on their activities. As another example, the interviewee cited in quote #6 reflects on community activities and highlights the benefits of a community leader who aligns the community towards a shared goal and vision.

Data furthermore indicates that community leaders that abandon from commands and orders in community settings are more successful in establishing a shared vision within the community. Quote #7 and #8 underline this argument. As the interviewees describe, community leaders should support the community in identifying and establishing their shared vision as opposed to dictate a desired output as well as the way in which this output is achieved. While it is beneficial to have a competent consultant in community processes, it is also important that this consultant does not act as a supervisor of other community members because this would be detrimental for cognitive community social capital. Related to that, quote #9 and #10 stress the differences between hierarchy and leadership in community settings. As the interviewees state, leadership within community settings should not reflect hierarchical structures in organizations.

Table 25: The role of community leadership

Illustrative quotes: Community leadership → Cognitive social capital
1. "A community needs someone who takes care of it [...]. It's better to manage and restrict the community to a certain extent." *Knowledge & idea manager (Case B: Automobile)*
2. "The most important thing is to allow the community to grow slowly. And if something develops badly, you need to recognize it and to steer into the opposite direction." *R&D manager (Case B: Automobile)*
3. "The one who leads the community should possess a certain appearance that makes you curious about community activities. He or she has to be able to activate potential community members and to strike the right note [...]. For instance, in community settings you don't need a manager that aims to preach boring information presented with overhead or PowerPoint slides." *Community member (Case C: Toy manufacturing)*
4. "The community leader provided us a summary of preliminary results of community activities from time to time and everybody agreed with that summary." *Community member (Case C: Toy manufacturing)*
5. "I think it was really beneficial to have someone who prepared the workshops in advance." *Community member (Case D: Automation)*
6. "Retrospectively, I realize that we aimed to achieve a certain goal during each community workshop. You have to have someone who is responsible to organize community innovation processes somehow." *Community member (Case E: Conglomerate)*
7. "In the community, everybody took responsibility of a certain task [...]. There was no need for anybody who assigns responsibility to others. It was a nice spirit because you didn't have to talk a lot about responsibilities." *Community member (Case C: Toy manufacturing)*
8. "One group had a mentor from the top management of our company [...]. He said to the group: 'This job is your issue. I do not tell you what to do. I am there for you if you need help or if you need a consultant. Otherwise, I will be invisible' [...]. As a result, the group identified heavily with their work." *General manager (Case B: Automobile)*
9. "I think it was important that the community leader was a neutral person – a supervisor or boss is not suitable for this job." *Community member (Case C: Toy manufacturing)*
10. "I think for community work, it is very important that the boss does not participate. You are able to exchange more openly as a result." *Community member (Case D: Automation)*

The literature provides related arguments that support my findings. The absence of hierarchy is often mentioned in the extant literature as a key characteristic of innovation communities (e.g., Fleming & Waguespack, 2007). Muhdi and Boutellier (2011), for instance, find that the minimized role of hierarchical aspects in internal innovation communities plays a central role with regard to motivation of community members to participate in innovation processes. Also related to my findings, Bono and Anderson (2005) study how transformational leadership (i.e., leadership that is characterized by charisma, intellectual stimulation, and individualized consideration) affects social capital. They find that transformational leadership is positively related to network centrality of leaders as well as their reports. Another related study has been published by Brockman et al. (2010) who find that formalization of interaction within a team is negatively associated with social cohesion because it inhibits interaction between team members. Similarly, Inkpen and Tsang (2005) argue that decentralized authority in networks allows members to use their knowledge best. Another related literature stream is concerned with discrepancies between individual leadership and corporate leadership systems. Huff and Möslein (2004) review extant literature on this topic and suggest that environmental developments have led to new distributed forms of peer- or shared leadership. These arguments are closely related to the forms of leadership required in corporate innovation communities.

5.4.3 Workshop and platform design

Beside task characteristics and community leadership, I identified design elements of community workshops as well as online platforms as antecedents of social capital in corporate innovation communities. Table 26 summarizes my findings with regard to workshop designs where I structure my arguments in three subcategories of design elements. First, quote #1 and #2 emphasize the importance to start community workshops with a short self-introduction of community members. By definition, such processes are associated with interaction between previously strangers and, hence, community members establish ties and structural social capital in such settings. At some community workshops I observed, community members used specific methodologies that forced each member to talk to all other participants of the workshop for a minute. The quotes cited indicate that the interviewees perceived this kind of introduction of participants as valuable. To sum up, I suggest that introduction of participants is related to structural social capital in corporate innovation communities.

Second, quote #3 highlights the importance of a climate that supports unconventional ideas and refrains from evaluating the ideas in the early stages of innovation processes during community workshops. In some communities I observed, community members agreed on specific rules for community work that support innovative collaboration. Such rules are beneficial for structural social capital as community members have the courage to exchange intensively on the ideas they develop in community settings. In addition, these rules support a shared vision of community work among members as workshop participants identify with the innovative working mood in community settings.

Third, my interviewees mentioned a number of methodologies they used during workshops that supported interaction (structural social capital) and a shared vision and language (cognitive social capital). For instance, quote #4 and #5 point to the importance of changing working groups within the community to countervail the concentration of interaction within small associations of community members. This is especially important because community members often belong to different functions and tend to concentrate interaction on peers who have a similar background (Chua et al., 2008). The interviewee cited in quote #6 refers to a specific methodology (i.e., LEGO Serious Play) that supported the establishment of network ties (i.e., structural social capital) in the community. Quote #7 of a member of the community at the toy manufacturing company explains why methods that involve playing are beneficial for structural social capital. As the interviewee describes, community members interact with many others in such situations and, hence, establish structural social capital. Observations at workshops showed that application of unconventional methods such as LEGO Serious Play was also associated with shared visions within the community because these methods differentiated community activities clearly from other corporate settings and supported identification with the community as a whole.

Table 26: The role of workshop design

Illustrative quotes: **Workshop design → Structural and cognitive social capital**
1. "The self-introduction of participants at the kick-off event was excellent. We introduced ourselves in a way I have nether witnessed before." *Community member (Case E: Conglomerate)*
2. "Of course, we also introduce participants at our company. However, this happens often in an old-fashioned way. Because we used other methodologies, there was a wind of change blowing within the community." *Community member (Case E: Conglomerate)*
3. "The innovation community is a playground. You are allowed to come up with weird ideas. For me, this is really important." *Community member (Case E: Conglomerate)*
4. "The assembly of community members during group work was really important. During workshops, we always worked in small groups – and the composition of group members changed constantly. This was beneficial, because people tend to build small sub-groups within collectives." *Community member (Case E: Conglomerate)*
5. "I think it was important to change members of subgroups within the community. Because of that, we didn't build frontiers within the community." *Community member (Case D: Automation)*
6. "Because we used methods that forced us to interact heavily during community workshops (e.g., LEGO Serious Play), we concocted a plan from the very beginning. The whole community used the same terms – even across different groups within the community." *Community member (Case A: Temporary employment)*
7. "The most intensive interaction took place while we were playing games during community workshops. I mean, the sense and purpose of games is to have fun together. Of course we were chatting a lot." *Community member (Case C: Toy manufacturing)*

In a nutshell, I propose that workshop designs that support interaction between community members (i.e., self-introduction of community members; rules that support exchange of unconventional ideas; methodologies that support interaction) affect structural as well as cognitive community social capital positively. Related arguments are put forward by Lawson, Petersen, Cousins, and Handfield (2009) who find that formal socialization mechanisms create opportunities to develop informal relationships. That is, community workshops that receive organizational support create opportunities to establish social capital, which can be used in subsequent settings. The study of Bolino et al. (2002) relates to my findings as well. Bolino et al. (2002) argue that participation in nonmandatory meetings, such as community workshops, is positively related to the formation of network ties. At the same time, this social participation helps to develop trust because it creates opportunities to interact with other community members socially. With regard to the cognitive dimension, Bolino et al. (2002) argue that voluntary participation in meetings helps employees to gather information about developments inside the organizations and, as a result, employees are able to understand other members of the organization better. In essence, the arguments of Bolino and colleagues (2002) support my suggestions that workshop designs that support interaction are positively related to structural community social capital, which in turn fosters trusted relationships as well as shared visions and languages inside communities. Related to the rules of behavior within community settings, Hughes and Perrons (2011) have advocated for the benefits of clearly defined norms in the creation of cognitive social capital. Schrage (2000) discusses the benefits of prototyping in innovation processes as prototypes allow individuals to discuss ideas intensively. In addition, prototypes enable community members to adapt their mental models of the innovation process based on the inputs of others. These arguments are closely related to my findings with regard to the benefits of LEGO Serious Play. In essence, LEGO Serious Play is often used in community workshops as a means to create prototypes on the basis of which community members discuss future activities. As Doll (2009) suggests, prototypes represent boundary objects that allow heterogeneous groups to exchange knowledge and ideas. As a result, such boundary objects support interaction as well as a shared vision of desired outcomes in innovation processes.

In addition to design elements of community workshops, my data suggests that online community platform designs also antecede structural and cognitive social capital. For instance, quote #1 in Table 27 describes general patterns of interaction on community platforms (e.g., searching for, selecting and implementing ideas), which allow community members to establish and strengthen relationships with other community members. In quote #2, the interviewee describes the importance of platform functionalities that support a virtual presentation of each community member in order to be able to identify and approach unknown community members who possess valuable resources.

However, many community members also expressed doubts whether interaction on community platforms would have succeeded if they had never met their peers in offline settings before[22]. For instance, interviewees cited in quote #3-#5 articulate their doubts that online interaction is a substitute for offline interaction in corporate innovation community settings. Examined through a social capital lens, the

[22] Of course, as there are numerous examples of successful communities operating exclusively in virtual settings (e.g., Faraj et al., 2011), qualitative evidence presented here should be treated with caution.

quotes cited indicate that structural social capital in online settings represents a weaker basis to establish relational and cognitive social capital.

Moreover, data indicated a pivotal role of user interfaces of community platforms in the establishment of social capital. Quote #6 describes the role of user interfaces in establishing cognitive social capital of community members. As stated, a user interface that represents the corporate identity of the company that supports community activities facilitates the development of a shared vision of community members. Data also points to the importance of usability of community platforms in order to strengthen interaction and structural social capital in community settings (quote #7). I observed this relationship also in community meetings based on virtual whiteboards. Those community members who had usability problems did not exchange ideas intensively with other community members during those meetings and were not able to build ties with many community members.

Table 27: The role of platform design

Illustrative quotes: **Community platform design → Structural & cognitive social capital**
1. "There is a company-internal platform that allows employees to present their ideas to others. [...]. Subsequently, other employees can study the ideas, can add comments, ideas can get actualized and evaluated." *Community member (Case D: Automation)*
2. "The platform provides an opportunity to upload a picture and a short text to describe oneself. Hence, it is pretty easy to contact other community members this way." *Knowledge & idea manager (Case B: Automobile)*
3. "I don't think that an online platform is able to substitute face-to-face communication. I think it strengthens communication [...]. That's my opinion on that." *General manager (Case B: Automobile)*
4. "Workshops in the headquarter are important to get to know each other. After that, you can work successfully in virtual meetings." *Community member (Case A: Temporary employment)*
5. "I am not sure whether the virtual meetings would have succeeded if we had no clue about the other participants." *Community member (Case A: Temporary employment)*
6. "For us, it is really important that the user interface represents our corporate identity [...] because user interfaces are able to create affinity. There is a signal: The platform is flawless and belongs to the company." *Knowledge & idea manager (Case B: Automobile)*
7. "I am convinced that our community platform is designed in a way that makes training and practice unnecessary. It's as easy as surfing the web with a standard browser." *Knowledge & idea manager (Case B: Automobile)*

I propose that community platform design is both related to structural as well as cognitive social capital. My findings are related to the arguments of many scholars who have emphasized the role of advances in ICT in the establishment of the paradigm of open innovation (Baldwin & von Hippel, 2011; Dahlander & Wallin,

2006). Many scholars have examined how community members rely on ICT in order to innovate (e.g., Stam, 2009). While some scholars argue that social relationships do not play a major role in online communities (e.g., Faraj et al., 2011), other studies highlight the role of social capital for knowledge contributions in such environments (e.g., McLure Wasko & Faraj, 2005). My findings relate to extant literature as I observed that online community platforms helped community members to sustain and expand social capital. However, interviewees also expressed serious doubts that it would have been possible to socialize with other community members unexceptionally in online settings and emphasized the importance of community workshops for the creation of social capital.

The role of user interfaces of community platforms is also discussed by extant researchers (e.g., Jung, Schneider, & Valacich, 2010). For instance, Ma and Agarwal (2007) find that different functionalities (e.g., the opportunity to provide detailed profile information) affect knowledge contribution in online settings[23]. Schröder and Hölzle (2010) argue that usability is positively related to interaction quality within communities. In other words, usability determines if and how intense community members make use of the online community platform (Nielsen, 2000).

To sum up, I identified workshop as well as community platform design as antecedents of social capital on the community level. With respect to workshop design, data indicated that self-introduction of participants, rules that support innovative exchange between members as well as interaction supporting methods antecede social capital in community settings. While interviewees also described community platforms as helpful to support interaction, my empirical observations also highlight that interaction on online platforms does not affect relational and cognitive social capital as heavily as offline interaction does. Figure 18 summarizes the findings of the framework on the community level.

[23] This relationship is mediated by perceived identity verification, which captures "the perceived confirmation from other community members of a focal person's belief about his identities" (Ma & Agarwal, 2007: 46).

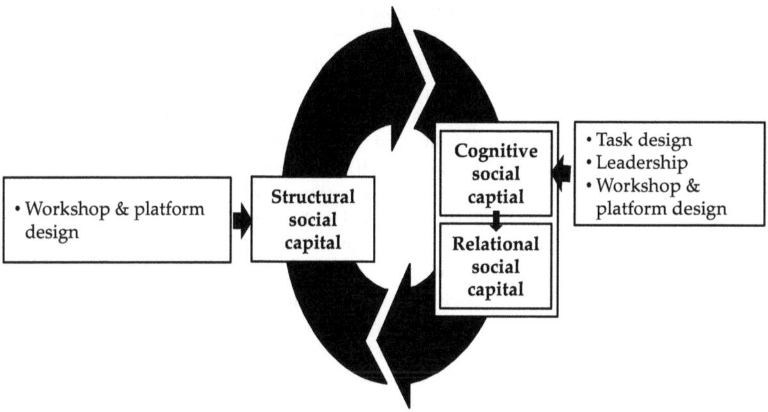

Figure 18: Community level antecedents of social capital

In a nutshell, the framework presented in this chapter offers insights on antecedents of social capital in corporate innovation communities on three interrelated levels. Table 28 summarizes my findings with regard to individual, organizational, and community level antecedents of corporate innovation community social capital.

Table 28: Multilevel antecedents of corporate innovation community social capital

Level	*Sub-Categories*	*Impact on Social Capital*	*Related Literature*
Individual	Personality	Openness to experience, agreeableness, emotional stablility, and extraversion affect structural social capital positively; conscientiousness is positively related to structural as well as cognitive social capital	Klein et al. (2004); Oh & Kilduff (2008)
	Motivation	Extrinsic motivation affects structural social capital positively and intrinsic motivation is associated with both structural as well as cognitive social capital	Nov (2007); Roberts et al. (2006)
Organizational	Organizational culture	Collaborative and innovative corporate climates are beneficial for both structural as well as cognitive social capital	Inkpen & Tsang (2005); Leana & van Buren (1999)
	Formal and informal organization	Formal and informal organization affect structural social capital	Brass et al. (2004); Gulati & Puranam (2009)
Community	Creative task design	There is an inverted U-shaped relationship between creative task design and cognitive social capital	Lakhani & Wolf (2005); Müller-Seitz & Reger (2009)
	Community leadership	Community leaders affect the cognitive dimension of social capital	Bono & Anderson (2005); Fleming & Waguespack (2007)
	Workshop & platform design	Workshop as well as platform designs affect the structural and the cognitive dimension of social capital	Doll (2009); Jung et al. (2010)

6 Discussion

Part III adds to research on innovation communities and social capital by offering a comprehensive framework that describes antecedents of corporate innovation community social capital. Extant research often examines consequences of social capital, but is lacking rigor with regard to antecedents of social capital (Payne et al., 2011). I discuss theoretical contributions, findings that conflict with extant publications as well as limitations of the study presented in the following[24].

The theoretical framework presented in this part entails two important aspects. First, the framework conceptualizes social capital as a multidimensional concept that reinforces itself and second, it categorizes antecedents of different dimensions of social capital on three levels. I discuss these two aspects of the framework in the following and close this part with raising issues of the framework that conflict with extant findings as well as limitations of the study presented.

6.1 Community social capital as a multidimensional phenomenon

A key contribution of the research presented in this part is the multidimensional nature of the theoretical framework offered. In more detail, the framework entails propositions with regard to structural, relational, and cognitive dimensions of social capital (Nahapiet & Ghoshal, 1998). Extant theory on innovation communities often highlights the importance of all dimensions of social capital in such settings (e.g., McLure Wasko & Faraj, 2005). In a similar vein, social capital scholars emphasize the importance of all three dimensions of social capital for innovative performance (Tsai & Ghoshal, 1998). Yet, to the best of my knowledge, research on antecedents of social capital in innovation communities neglected the interrelations between different social capital dimensions largely.

In this context, I propose that the three dimensions of social capital reinforce each other. In more detail, the framework presented suggests that structural social

[24] At the end of the dissertation, part V offers a general discussion, including theoretical and managerial implications that result out of a simultaneous consideration of the studies presented in part III as well as part IV.

capital antecedes and is positively affected by cognitive and relational social capital. For example, I have argued that individuals who exchange knowledge in community settings create social relationships and, thus, the structure of the social network within the community changes. While exchanging knowledge within the community, community members have the opportunity to foster the quality of relationships (e.g., trust) within the community (Nahapiet & Ghoshal, 1998). At the same time, the cognitive dimension of social capital is affected as community members develop a shared vision and language while working together. In turn, trust towards the community as well as a shared vision and language prompts community members to build additional ties within the community. I propose that my conceptualizing of social capital as a reinforcing cycle of interrelated dimensions adds to the arguments of other scholars who have advocated for relationships between different dimensions of social capital (e.g., Tsai & Ghoshal, 1998).

6.2 Community social capital antecedents on multiple levels

A second contribution I offer in this part of the dissertation is the three different levels of social capital antecedents in corporate innovation community settings. In my framework, I have conceptualized those antecedents on the individual level that affect social capital of single individuals. On this level, data indicated that personalities as well as extrinsic and intrinsic motivations of individual community members antecede social capital. In contrast to the individual level category, my category of organizational level antecedents captures antecedents that affect the organization, which supports innovative activities, as a whole. The supporting organization provides the environment in which community social capital grows. In this category, I have identified an organizational culture that promotes collaboration and innovation as well as aspects of formal and informal organization as relevant antecedents.

Antecedents on the individual and on the organizational level determine community social capital to a significant extent. For instance, innovation communities benefit from community members that are intrinsically motivated because these individuals are likely to develop shared visions with other community members. At the same time, an innovation community benefits from a facilitating organizational culture for community activities. A culture that promotes mutual assistance, collaboration and innovative activities promotes structural as well as cognitive social capital in the organization and the community.

In addition, I have argued that community level social capital antecedents play a crucial role in establishing social relationships as well as shared visions and languages among community members. My category of community level antecedents refers to those antecedents that affect the community as a collective. In contrast to organizational level antecedents, employees who do not participate in community activities are not affected by this category. On the community level, data indicated that creative task design, community leadership as well as workshop and platform design are important antecedents of community social capital. While many of the relationships proposed in the framework have already been mentioned in previous studies, the study presented in this part of the dissertation adds to the literature in two ways. First, I empirically examine whether theoretical arguments established in other settings are tranferable to innovation community settings. Second, a key contribution of the framework offered is its comprehensive analysis of antecedents of social capital that acknowledges both the multiple dimensions and multiple levels of social capital.

6.3 Differences to extant studies on antecedents of social capital

While the framework offered is consistent with most of the arguments put forward by other scholars who look at antecedents of social capital, I deviate from the findings of the literature with respect to three important aspects. First, scholars have argued for a tendency of individuals to approach others that possess similar characteristics (e.g., Chung, Singh, & Lee, 2000). For instance, Carpenter et al. (2012) suggest that homophily of actors determines formation of social ties to a significant extent. As another example, Grossman, Yli-Renko and Janakiraman (2012) find a moderation affect of age based and gender based similarity on the relationship between resource multiplexity (i.e., the number of resources available through a focal contact) and perceived value of the focal contact from the perspective of a nascent entrepereneur. Vissa (2011) studies antecedents of intentions to form ties of Indian entrepreneurs and finds that being from the same castes and speaking the same language are significant antecedents of intentions to form ties.

I did not observe a tendency of community members to prefer social relationships with similar others (e.g., with respect to age or education). While community members tended to interact with peers who they already knew because of formal organization (an aspect related to functional similarity) at the beginning of a community project, I observed that community members liked interaction with peers who were able to provide different perspectives on community work at the same time. In this context, I suggest that three differences between community activities and other empirical settings are important for the distinctness of the framework proposed and extant studies. First, those who decide to engage in community activities look for environments that oppose them with heterogeneous peers in order to support innovative knowledge and idea combination. Second, workshop designs worked against the formation of subgoups within the community and did not allow members to follow their unconscious tendency to approach similar others, for example via changing subgroups within the community. Third, all community members share similarities with regard to their interest in the supporting company and innovation processes. Hence, this similarity across all community members may be sufficient to stimulate interaction.

Second, many scholars have argued that access to resources of an individual is an important antecedent of social capital (e.g., Hughes & Perrons, 2011). Cowan and Jonard (2009) present a model that conceptualizes knowledge complementaries between partners in innovation alliances as the most important antecedent of network structures. Relatedly, Lee (2010) finds that individuals who have demonstrated the value of their resources and capabilities are able to form more non-redundant ties. As another example, Baum, Cowan, and Jonard (2010) examine how aspects of knowledge similarity affect selection of network partners in innovation networks. In their argument, they highlight that partner companies in innovation networks need to have moderate levels of knowledge overlaps as too much overlaps hinder innovative outcomes and too few overlaps hinder understanding each other.

Data gathered does not suggest that resource complementarity constitutes a major antecedent of corporate innovation community social capital. In contrast, interviewees often stressed that each community member contributed to the accomplishment of the community task. I argue that this finding is related to the motivation of community members to work together in a joint effort in innovation processes (Franke & Shah, 2003) instead of focusing interaction on single community members.

Third, and conflicting with other researchers, I did not find evidence for antecedents that affect the relational dimension of social capital directly. Data gathered indicated that trust developed because of interaction or shared visions in community settings. In contrast, Blatt (2009) suggests that written or oral contracts between entrepreneurial team members are positively associated with trust because contracts make behavior more predictable. Blatt's (2009) findings are closely related to my arguments for rules that encourage idea exchange in community settings. However, the framework offered in this part argues that these rules do not affect trust directly but through interaction within community settings. Again, I suggest that differences between earlier research settings and my findings are attributable to the specific situation of community settings. Obviously, and in contrast to contracts signed in other settings, the rules established in community workshops do not cause an entitlement by law. As a result, I argue that community members do not believe that each community member will follow the rules they have agreed upon. However, the rules encourage community members to interact intensively and develop trust as a result. In hierarchy-free environments, formal arrangements or rules are not suitable to create close and trusted networks. In contrast, trust emerges as community members exchange and elaborate on joint ideas.

6.4 Limitations and avenues for future research

Although promising, the study presented in this part has a number of limitations. I present my most important concerns in this section. My first key issue is related to characteristics of the data gathered. In three of the five cases analysed (Cases A, C and D), I accompanied the establishment of the corporate innovation community from a scientific point of view. In this context, I supported the company in the preparation and moderation of community workshops. One could argue that the research design applied envolves aspects of action research (Baburoglu & Ravn, 1992; Lewin, 1947) as a result. Although my activities were not directly related to facets of social capital and

I was only one of many other researchers and moderators involved, there remains a risk that my activities have biased the findings presented. I call for further studies in settings where researchers did not influence community processes at all.

A second limitation of my sample is associated with community characteristics. Three of the communities under research allowed external members to participate. However, in these communities, external members had worked for the company in the past and retired a few months or years ago. The reason behind this procedure is that these organizations used innovation communities to facilitate innovation processes with age diverse groups. Although the data analysed suggests that the findings are transferable to other contexts, future studies are needed in this area. For example, are there greater effects of nondisclosure if community members were never employed by the company? While nondisclosure was also relevant in the cases analysed in this part of the dissertation, it may be more pivotal in settings where external community members never had a legal relationship with the supporting organization.

Third, I call for future research endeavours that test the propositions I have stated in this part of my dissertation. While a qualitative research approach is helpful to advance scientific knowledge in an area where extant understanding is limited (Eisenhardt, 1989), it is interesting to complement it by quantitative approaches that are able to test the significance of proposed relationships. For many of the contructs involved in my propositions, measurement scales are readily available in the literature. Thus, it is possible to test the propositions stated in this part of the dissertation. Because the framework offered involves a number of latent variables and mediated relationships between antecedents and self-reinforcing dimensions of social capital, I argue that SEM provides an appropriate research design. In the next part, I take a first step and test my propositions with respect to the individual level antecedents of social capital, applying CBSEM.

IV

Individual level antecedents of social capital in corporate innovation communities:

A SEM approach

1 Structure of part IV

Part IV builds on the previous part III and focuses on the individual level of community social capital. The model presented here takes a look at personality traits (i.e., the Five Factor Model [FFM]) (Goldberg, 1992) as well as intrinsic and extrinsic motivations (Deci, 1972) to engage in community activities as antecedents of social capital in corporate innovation community settings. By offering more nuanced theoretical lines of arguments as well as quantitative empirical evidence with regard to the relationships between personality as well as motivations of community members and social capital, part IV presents interesting insights for both researchers as well as practitioners attached to recruiting and incentivicing community participation. While part III addressed community management as a whole, this part focuses on consequences of community members' personality and motivations with regard to social capital.

The empirical field to test the hypotheses derived out of qualitative research efforts discussed in part III as well as theoretical considerations is an innovation community of bachelor students at FAU. In contrast to the previous part, I take a quantitative research approach. The model presented considers different dimensions of social capital and takes account of the network size of community members (structural dimension), of shared visions and languages (cognitive dimension), and of the quality of the relationships between community members in terms of trust as well as tie strength (relational dimension). In line with qualitative evidence derived in part III of this dissertation, I argue that different dimensions of social capital are interrelated and reinforce each other.

Using CBSEM (AMOS 20.0.0), I find support for most of my hypotheses. In more detail, I find empirical evidence that network size is positively related to shared visions and languages, to tie strength as well as to trust towards other community members. Shared visions and languages are positively related to tie strength as well as trust towards other community members. In addition, I find that tie strength is positively related to trust towards network ties in the corporate innovation community. Furthermore, the personality trait of emotional stability is positively related to network size of community members, while introversion is negatively related to network size. Intrinsic motivation and conscientiousness are positively related to shared visions and languages within corporate innovation communities. Out of these empirical findings, I conclude that personality and intrinsic motivation are significant antecedents of innovation community social capital, while extrinsic motivation is not. In other words, personality traits of community members significantly affect the extent to which a focal individual exchanges and develops shared visions with other community members. At the same time, intrinsic motivation of community members affects the extent to which community members develop a shared vision within the community. In contrast, the role of extrinsic motivation as an antecedent of social capital in innovation community settings was not confirmed.

Part IV is structured similar to the stage process for SEM proposed by Hair, Black, Babin, and Anderson (2010). In the *first stage*, I review relevant literature for individual level antecedents of social capital, i.e., personality traits as well as extrinsic and intrinsic motivations. Qualitative data gathered at five different innovation communities indicated that these constructs affect the extant to which individuals build social capital in community settings. Concerning personality traits, I focus on the Five Factor Model (FFM). The main premise of the FFM is the identification of five personality traits, which "capture the essence of one's personality" (McElroy, Hendrickson, Townsend, & DeMarie, 2007: 810) and provide "a useful set of very broad dimensions that characterize individual differences" (Digman, 1990: 436). While researchers have labeled these "Big 5" factors differently, the labels most often used nowadays are (1) openness to experience, (2) agreeableness, (3) emotional stability, (4) introversion, and (5) conscientiousness (e.g., Wolff & Kim, 2012). In addition to the literature on personality traits, I discuss extant research on intrinsic and extrinsic motivations in innovation communities (e.g., McLure Wasko & Faraj, 2005).

In *stage two*, I develop the theoretical model. McDonald and Ho "simply suggest what should be obvious, namely, that the theoretical justification of directed arcs should be given as much detail as possible, hence open to the plausibility judgment of a reader with relevant substantive expertise" (2002: 67). This section responds to their recommendation, treats qualitative findings of part III as a starting point, and develops detailed theoretical lines of argument to underline hypothesized relationships.

Stage three describes the empirical setting. Data was gathered from members of the WIN Innovation Community, a contest-based innovation community that is organized every year in the course of a bachelor lecture at the FAU. Students within the community focus on different innovative tasks. Community activities take place in both offline settings (e.g., team meetings) as well as on online platforms. While participation at the WIN Innovation Community in the sense of submitting a final concept is mandatory to the students, the extent to which students exchange ideas with community members and others outside their team remains voluntary to a large extent. Moreover, the students within the community are not hierarchically organized and receive support of the university in developing their concepts. Therefore, the WIN Innovation Community meets the definition of an innovation community provided in part II.

In *stage four*, I present my approach to data analysis, give a short introduction on SEM in general and discuss advantages and disadvantages of the focal research method. Structural equations modeling is a suitable research method in the context of the study presented, most importantly because it allows a researcher to deal with latent variables as well as to study mediator variables (Diamantopoulos & Siguaw, 2000; Gefen et al., 2011). I elaborate on exploratory (EFA) as well as confirmatory factor analyses (CFA) that are also relevant in the context of SEM. Different indices used in this study to examine the fit of both my measurement and structural model are discussed, too.

Stage five presents exogeneous and endogeneous measures of the model. It provides a review of extant literature and presents scales used in prior studies for both exogeneous and endogeneous variables. I discuss the procedure in pretesting the measures and show how scales were finally adopted to the present context. Subsequently, I apply the procedures of EFA and CFA and take a look at model fit indices in order to examine measurement model validity.

In *stage six*, I discuss the model fit of the overall model, test my hypotheses and take a look at multiple squared correlations. In addition, I consider alternative models in order to verify the robustness of my findings. As I have already mentioned, data confirms most of my hypotheses.

In *stage seven*, I discuss theoretical implications and limitations of my findings. The theoretical contribution of the study presented in this part is threefold. First, the research presented adds to the literature on social capital and demonstrates how personality and intrinsic motivation to engage in community activities are significant antecedents of innovation community social capital. Second, my findings contribute to a more comprehensive view of social capital and its antecedents as the model considers interrelationships between four constructs that represent different dimensions of social capital. Third, I add to the literature on innovation communities. More specifically, integrating social capital theory allows me to examine aspects of extrinsic and intrinsic motivations in a more detailed way. I find that extrinsic motivation does not affect social capital. In other words, extrinsic motivation does not lead to the formation of ties to other community members. To the contrary, intrinsic motivation is associated with cognitive social capital (i.e., shared visions and languages) which, in turn, is associated with strong ties and trustful interaction within the community. The structure of part IV is summarized in Figure 19.

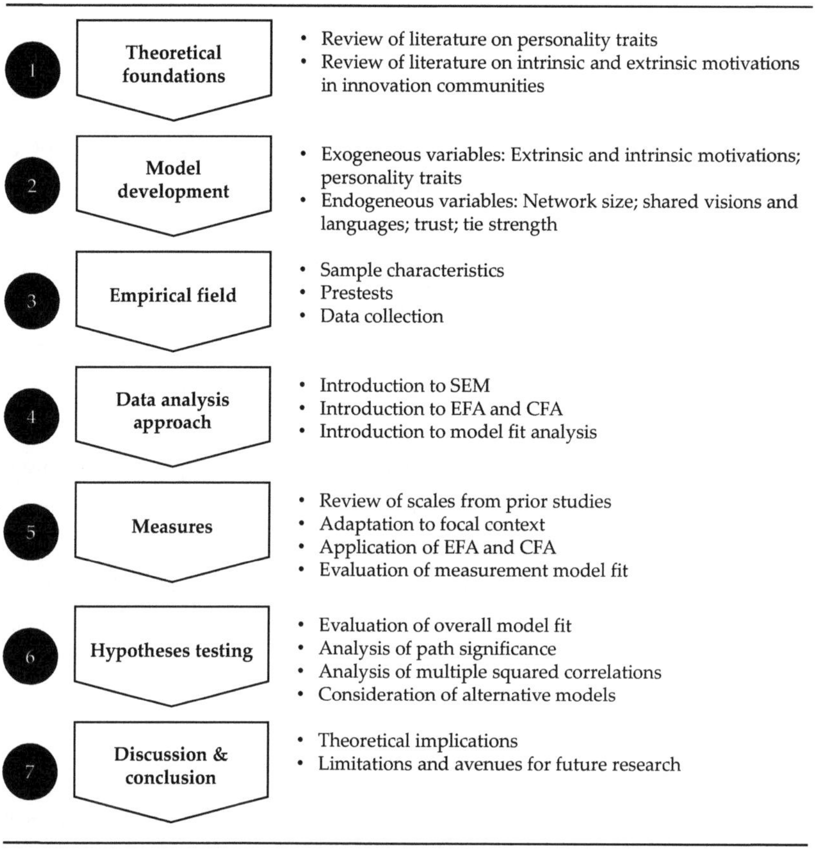

Figure 19: Structure of part IV based on Hair et al. (2010)

2 Theoretical foundations

In this chapter, I discuss the theoretical foundation of the exogeneous variables of the model. As qualitative data presented in part III indicated, personality is related to structural as well as cognitive social capital and affects the extent to which individuals develop ties and shared visions with other community members. At the same time, qualitative evidence suggested that motivation of community members to engage voluntarily in community activities affects the number of ties as well as the development of shared visions and languages with other community members.

I start by reviewing the literature on personality traits and focus on the most prominent Five Factor Model (FFM) (Goldberg, 1992) and its application in studies related to management and business administration. Thereafter, I give an overview of studies that look at motivations of participants in innovation communities. The theoretical background of the endogeneous variables (i.e., structural, relational, and cognitive community social capital), is not discussed as it is already presented in part II.

2.1 Personality traits

Personality is a major determinant of human behavior as it "reflects the unique facets of each human being, the traits that define our essence, and it is reflected in all of our thoughts and actions" (Devaraj, Easley, & Crant, 2008: 94). As such, personality is an important contruct in explaining economic phenomena and many researchers have looked at the consequences of different personality traits in managerial and business contexts. For instance, Villani (1975) already examined many years ago the relationship between personality and the type of television programme an individual watches – a question highly relevant for marketing activities of companies. In a similar vein, Alpert (1972) looks at the relationship between personality traits and product choice, i.e., place of residence. He finds that personality profiles have a significant influence on the choice of place of residence. For instance, individuals who describe themselves as casual and free-spirited choose residents with a friendly atmosphere, large rooms and unclean appearance. To the contrary, bachelor students who are self-centered and value seclusion live in residents that provide privacy and an unfriendly atmosphere (Alpert, 1972). Other scholars take the perspective of a HR department and study how a number of personality traits affect job performance (e.g., Ashton, 1998). For instance, Pruden and Peterson (1971) find that the personality trait of power is positively related to job performance of salesmen. More recently, Camgoz, Karan, and Ergeneli (2011) investigated the relationship between personality traits and job performance of fund managers. Camgoz et al. find that the trait of extraversion is positively related with financial success of fund managers because extraverted people "seek [...] out activity, dominance and stimulation of highly dynamic action environments" (2011: 148). In addition, they find that the personality trait of conscientiousness is positively related to the performance of fund managers as this trait is associated with discipline and self-control.

In a nutshell, personality is an important construct in managerial and business contexts. However, studying issues related to personality is far from trivial and the literature witnesses a long and intense discussion on how to conceptualize personality and its traits (Digman, 1990). Within this long lasting academic debate, scholars have proposed different categorizations of personality traits. For instance, the Edwards Personal Preference Schedule distinguishes between 15 personality traits (i.e, achievement, deference, order, exhibition, autonomy, affiliation, intraception, succorance, dominance, abasement, nurturance, change, endurance, heterosexuality, and aggression) (Edwards, 1954) and is often used in the field of marketing studies (e.g., Alpert, 1972; Brody & Cunningham, 1968). As another example, Pruden and Peterson (1971) have relied upon self-other orientation theory to investigate personality issues. Self-other orientation theory illuminates ones personality from the perspective of a significant other (e.g., friends or coworkers) and distinguishes between four personality traits: power, marginality, identification intensity, and openness (e.g., Pruden & Peterson, 1971).

Nowadays, the Five Factor Model (FFM) is the most popular conceptualization of personality (Goldberg, 1992). The major proposition of the FFM is that personality can be described by five distinct personality traits, also referred to as the "Big 5". These five factors have been identified via factor analytic procedures of numerous personality traits in large scale settings (Goldberg, 1992). It is important to note that although the FFM has received much attention and acceptance among scholars in the field of personality psychology (e.g., McCrae & Costa Jr., 1987, 1989) and management and business studies (e.g., Brown et al., 2002; McElroy et al., 2007), others have critized the framework, for instance because of the number of factors included (e.g., Hough, 1992). In the next section, I discuss the FFM and each of the "Big 5" traits in detail. In doing so, I rely on literature reviews and meta-analytic studies provided by others (e.g., Barrick, Mount, & Judge, 2001; Judge, Heller, & Mount, 2002) and complement their arguments with studies of particular relevance in the context of social capital and innovation communities.

2.1.1 The Five Factor Model of personality

The major proposition of the FFM is that five personality traits provide a basic structure of a trait taxonomy (McCrae & Costa Jr., 1987). Although researchers differ in the names of each trait as well as the precise definition, most scholars agree that personality consists of the traits of (1) openness to experience, (2) agreeableness, (3) emotional stability, (4) introversion, and (5) conscientiousness – often referred to as the "Big 5" (Barrick et al., 2001). In essence, advocates of the model argue that the FFM categorizes personality into "five broad factors and, as such, presents a concise yet comprehensive framework for studying personality" (Devaraj et al., 2008: 94). Mowen and Spears describe the "Big 5" as cardinal traits which consitute "the basic, underlying predispositions of individuals that arise from genetics and their early learning history" (1999: 410). Related to that, Gosling, Rentfrow, and Swann (2003) stress the high abstraction level of the FFM of personality and argue that each of the "Big 5" dimensions consists of a number of more specific traits.

The roots of the "Big 5" taxonomy go back to the study of Tupes and Christal (1961)[25]. However, these authors received marginal attention in the academic literature as they published their article in a technical journal of the United States Air Force (Barrick & Mount, 1991; Digman, 1990). As a result, Norman's (1963) replication study is often referred to as the origin of the FFM in the scientific literature. Since then, researchers have provided evidence for the validity of the FFM across cultures (McCrae & Costa, 1997) and measurement procedures (McCrae & Costa Jr., 1987).

[25] For a detailed discussion of the history on research on the FFM see Goldberg (1993).

There are numerous studies providing evidence that the "Big 5" significantly affect economic activities. In the context of innovation communities and antecedents of social capital, three types of studies are of particular relevance as they provide theoretical rationales for a relationship between the "Big 5" factors and social capital in innovation community settings. First, a number of scholars have investigated the relationship between the "Big 5" and job performance, often with a specific focus on teamwork or interpersonal tasks (e.g., Mount, Barrick, & Stewart, 1998). These studies are relevant for my dissertation because innovation community activities are by definition characterized by intense interpersonal interaction (e.g., Franke & Shah, 2003). Second, there are some studies that examine the relationship between the "Big 5" and technology acceptance (e.g., McElroy et al., 2007). This aspect is highly relevant in innovation communities because community members often rely heavily on IT to exchange ideas and knowledge with others (e.g., Bansemir et al., 2012). Third, extant scholars have examined the relationship between the "Big 5" and networking behavior (e.g., Wolff & Kim, 2012), i.e., the building and usage of social relationships. Obviously, the extent to which community members engage in networking activities within the community are relevant in studying antecedents of social capital. I review extant findings with respect to each personality trait in the following. Table 29 summarizes my arguments.

Table 29: The Five Factor Model of personality and relevant findings

FFM	*Other labels*	*Summary of relevant findings*	*Exemplary publications*
Openness to experience	Culture or intellect	(a) Conflicting findings with regard to teamwork; (b) positively related to technology acceptance; (c) positively related to networking	Jugde et al. (1999); McElroy et al. (2007); Wolff & Kim (2012)
Agreeableness	Friendliness	(a) Conflicting findings with regard to teamwork; (b) positively related to technology acceptance; (c) conflicting findings with regard to networking	Barrick & Mount (1991); Devaraj et al. (2008); Wanberg et al. (2000)
Emotional stability	Neuroticism	(a) Positively related to performance in teamwork; (b) positively related to technology acceptance; (c) no empirical support for relationship to networking	McElroy et al. (2007); Mount et al. (1998); Asendorpf & Wilpers (1998)
Introversion	Extraversion	(a) Negatively related to performance in teamwork; (b) negatively related to technology acceptance; (c) negatively related to networking	Barrick et al. (2001); Devaraj et al. (2001); Wolff & Kim (2012)
Conscient-iousness	Conformity or will to achieve	(a) Positively related to performance in teamwork; (b) positively related to technology acceptance; (c) conflicting findings with regard to networking	Devaraj et al. (2001); Mount & Barrick (1998); Wanberg et al. (2000)

2.1.1.1 Openness to experience

First, the factor openness to experience has divergent interpretations and meanings and researchers have referred to it as a double-edged sword (Judge et al., 2002). The factor has also been labeled by terms such as culture or intellect (Barrick & Mount, 1991). Some scholars argue that this factor of the FFM is the least understood trait in the model (e.g., Holland & Roisman, 2008).

The trait of openness to experience can be subdivided into the two characteristics of (1) intellectance and (2) unconventionality (Judge et al., 1999). First, the subdimension intellectance captures personality traits such as intelligence and broad-mindedness (Barrick & Mount, 1991; Barrick et al., 2001). Saucier (1994) attests individuals who are open to experiences the characteristic of philosophicalness. In addition, many researchers have described openness to experience by the terms cultured (Barrick & Mount, 1991; Mount et al., 1998) and intellectual curiousity (Camgoz et al., 2011; Wanberg, Kanfer, & Banas, 2000). Related to that, Devaraj et al. (2008) attest open individuals the ability to think flexible. Second, with regard to the unconventionality subdimension, Judge, Heller, and Mount (2002) argue that openness to experience envolves creativity both in the fields of science and arts. People who are open to new experiences are also imaginative and sensitive for new ideas (Barrick & Mount, 1991). McCrae and Costa (1987) argue that open people prefer variety instead of routines and describe them as untraditional. The factor of openness to experiences is also related to the traits of innovativeness (Saucier, 1994) and originality (Brown et al., 2002). Openness to experience entails a person's "willingness to explore new ideas" (McElroy et al., 2007: 811) and to use new techniques, methodologies and ways to reach goals (Wolff & Kim, 2012). As Camgoz et al. recently put it: "Openness to experience relates to [...] willingness to try different activities or consider unconventional ideas" (2011: 139).

I review the three research strands I have mentioned above that are relevant for studying openness to experience as an antecedent of innovation community social capital in the following. First, numerous scholars have explored the relationship between openness to experience and performance in different jobs and have produced conflicting results and theoretical lines of argument. Mount and others (1998) find that people who score high on openness to experience receive better ratings for their job performance from supervisors both with regard to teamwork as well as in service settings with customers. In a similar vein, Barrick and Mount (1991) find that openness to experience is positively related to training proficiency because the factor entails a motivation to learn. However, other studies do not find empirical support for a relationship between openness to experience and job performance. For instance, Camgoz et al. (2011) find no empirical support for their proposition that people who score high on openness to experience exhibit high performance as traders in financial market because they are less risk averse. Related to that, Barrick, Mount, and Judge (2001) find empirical evidence that openness to experience is not an important predictor of work performance. Judge et al. (1999) argue that openness to experience may be beneficial as well as unrewarding in the context of work performance. While openness to experience helps people in jobs where flexibility and creativity is important, it can be harmful in conventional jobs. In line with their arguments, Judge et al. (1999) find that openness to experience does not affect job performance significantly, but is negatively related to gravitation to conventional jobs.

Second, openness to experience explains the likelihood that a focal personality is going to accept and use a new technology (McElroy et al., 2007). Devaraj et al. propose that "those individuals high in openness are more likely to hold positive attitudes and cognitions toward accepting job-related technology [...]; they are less threatened by the change implied in adopting technology. This burning desire to do varied things that is the core facet of openness to experience will influence their judgment about the utility of the technology" (2008: 97). Devaraj et al. (2008) find that openness to experience affects the actual use of a new technology.

Third, scholars have investigated the relationship between openness to experience and networking behavior. For instance, Wanberg et al. (2000) argue that unemployed people who are more open engage more heavily in networking activities to find a new job because they are willing to try new and unconventional techniques to reach their goals. However, they find only marginal empirical support for their hypothesis. More recently, Wolff and Kim (2012) have argued that openness to experience is related to networking behavior of managers and employees because the two constructs share an informal focus. More specifically, open individuals engage in networking behavior because they are looking for new ideas. In line with their theorizing, Wolff and Kim (2012) find a positive relationship between openness to experience and networking behavior.

2.1.1.2 Agreeableness

Agreeableness or, as others have coined the factor, friendliness, describes whether an individual acts according to the interests of others (Camgoz et al., 2011). Likewise, Wolff and Kim propose that agreeableness contains "the mode of relating to others" (2012: 48). Nowadays, there is consensus among scholars that agreeableness entails two facets. First, agreeable people are described as being cooperative and caring about the well-being of others (Judge et al., 1999). In describing the essence of cooperativeness, scholars have used different terms. For instance, Barrick et al. (2001) highlight the trustful nature of agreeable people and argue that agreeableness can be defined by characteristics such as compliance. Others propose that cooperativeness of agreeable people also means that they react flexible in order to help others (Barrick & Mount, 1991). Related to that, McCrae and Costa Jr. (1987) stress that agreeableness is related to helpfulness and straightforwardness. In a similar vein, others characterize agreeable people as forgiving (McElroy et al., 2007) or thoughtful and supportive (Holland & Roisman, 2008).

Beside the cooperative nature of agreeable people, agreeableness captures the personal characteristic of likeableness which is reflected by good-natured, cheerful and gentle personalities (Judge et al., 1999). Judge and colleagues attest agreeable people a "motivation to achieve interpersonal intimacy" (2002: 531). Others describe agreeable people as affable (Barrick et al., 2001), courteous, soft-hearted or tolerant (Barrick & Mount, 1991). Likewise, Saucier (1994) characterizes agreeable people as kind and McElroy et al. (2007) describe them as sympathetic. Wanberg and colleagues (2000) argue that agreeableness entails aspects such as empathy and Devaraj et al. (2008) argue that agreeable people are considerate. In other words, agreeableness can be defined as the "warmth of feelings toward others" (Brown et al., 2002: 112).

Again, I review the three literature streams with respect to agreeableness I have mentioned above that are of particular relevance in a study on antecedents of innovation community social capital. First, the findings in the literature with regard to the relationship between agreeableness and job performance appear vague to a certain extent. For instance, Judge et al. state that "it certainly seems *possible* that [...] [agreeableness is] related to career success" (1999: 625). These authors argue that agreeableness has both positive as well as negative effects on career success. In situations where teamwork is crucial, the cooperative nature of agreeable persons may be beneficial. However, agreeable individuals may also fail to achieve their goals because being pleasant to others may be more important to them then pursuing their own or their bosses' interests. For instance, Brown et al. (2002) find that agreeableness is positively related to customer orientation of frontline employees in restaurants. Yet, agreeableness is negatively related to supervisors' performance ratings because agreeability "often stands in the way of productivity" (Brown et al., 2002: 115). In their empirical study, Judge et al. (1999) find that agreeableness negatively predicts extrinsic career success (measured by annual pretax income and occupational status). In a similar vein, Barrick and colleagues (2001) conclude from a meta-analytic study that agreeableness is a significant predictor of job performance in specific settings but that it is not able to predict overall job performance. More specifically, they find evidence that agreeable people perform in jobs that involve teamwork (i.e., interpersonal interaction, helping and nurting others, and cooperation). Likewise, Mount et al. (1998) study the influence of agreeableness on jobs involving personal interaction and find that agreeableness is positively related to performance in such settings, especially with regard to interactions in teams as opposed to dyadic relationships. To the contrary, also taking a meta-analytic approach, Barrick and Mount (1991) find only little support for their hypothesis that agreeableness is positively associated with performace in jobs that involve interpersonal skills (i.e., managers and employees working in sales departments). They conclude from their analysis that agreeableness does not contribute much to the explanation of job performance – neither with regard to overall job performance nor in specific settings that involve interpersonal interaction.

Second, scholars have also investigated the role of agreeableness in the acceptance and adoption of new technologies such as the internet (e.g., McElroy et al., 2007). Devaraj et al. (2008) find that agreeableness affects the use of new technologies in two ways. First, many new technologies allow people to cooperate with others easily. Because of the cooperative nature of agreeable people, they develop more positive believes about the usefulness of new technologies. Second, agreeable people also place value on the way others see them. Because of that, agreeable people will adopt a new technology more likely, especially if there is a subjective norm in their environment that demands them to do so.

Third, with regard to networking behavior, Wanberg et al. (2000) find that unemployed job seekers who score high on agreeableness make more intensively use of networking activities in order to find a new job. The authors argue that the cooperative and good-natured personality as well as their tendency to establish warm relationships with others fosters networking activities in job searching efforts. Wolff and Kim (2012) argue that agreeable people are more successful in networking because their cooperative and trustful nature reduces transaction cost with network ties. It is interesting to note that Wolff and Kim (2012) find no support for their hypothesis that agreeableness is positively related to networking behavior in general. Taking a more nuanced view, they find that agreeableness is more positively related to networking within the organization as opposed to networking across organizational boundaries. In addition, agreeable people spend more efforts in maintaining and using contacts as opposed to building new contacts.

2.1.1.3 Emotional stability

Emotional stability belongs, along with introversion, to the most often discussed and most pervasive personality traits (Barrick & Mount, 1991; Judge et al., 1999). Many researchers also labeled it by the other end of its continuum and referred to the trait as neuroticism (Judge et al., 2002, 1999; Wanberg et al., 2000). Emotional stability "captures the evenness or steadiness of a person's general emotional makeup" (Brown et al., 2002: 112) and belongs to the feeling related domains of personality traits (Wolff & Kim, 2012). People low on emotional stability or, vice versa, high on neuroticism, are often characterized by general negative attitudes (Judge et al., 2002) and lack the ability to order and organize psychological affairs (McElroy et al., 2007). The factor of emotional stability can be subdivided into two aspects.

First, neuroticism entails aspects related to anxiety (Holland & Roisman, 2008). For instance, Camgoz and others (2011) attest neurotic individuals a tendency to look for danger as well as imagination of bad aspects. Others describe individuals low on emotional stability as being overchallenged with stressful situations (McElroy et al., 2007). For instance, Mount and co-authors (1998) argue that neurotic people are bad in managing situations associated with physical or mental pressure. Related to that, scholars have also proposed that neuroticism captures characteristics such as impulsiveness (Judge et al., 1999) and vulnerability (Chen, 2011).

Second, emotional stability captures a person's well-being (Judge et al., 1999). For example, Holland and Roisman (2008) describe neurotic individuals as insecure, while Brown and co-authors (2002) assign them the characteristic of moodiness. Scholars also have characterized neurotic people as being angry rather than calm as well as by a lack of self-confidence and coolness (Camgoz et al., 2011). Others argue that neurotic people often are scarred, sad, embarrassed (McElroy et al., 2007) and easily upset (Gosling et al., 2003). Barrick and Mount (1991) propose that individuals with low emotional stability are often worried and depressed. Related to that, Saucier (1994) describes neurotic people as irritable and self-pitying. To the contrary, Mount and colleagues (1998) mark emotional stable people as relaxed, secure, and self-sufficient. In the following, I discuss findings of extant studies with regard to the relationships between emotional stability and job performance, technology acceptance, and networking behavior.

First, extant studies that investigate the relationship between emotional stability or neuroticism and job performance present contradictory findings. For instance, Camgoz et al. (2011) focus on job performance of fund managers and hypothize that emotional stability is associated with better decisions in the context of financial markets. However, they find no empirical support for this hypothesis. In a similar vein, Barrick and Mount (1991) expected a positive influence of emotional stability and job performance over a number of performance criteria as well as occupations. As was the case in the study of Camgoz et al. (2011), they find no evidence for their theorizing. In contrast, Judge et al. (1999) find that emotional stability is positively related to salary, number of promotions, and occupational status. Judge et al. (2002) find that emotional stability has a positive influence on job satisfaction.

Those scholars who focus on jobs related to teamwork and interpersonal affairs present findings that are more consistent. As an example, Mount et al. (1998) focus on jobs involving interpersonal exchange and find that emotional stability affects job performance in such settings. More specifically, they find that while emotional stability is a significant predictor of job performance in both teamwork as well as dyadic service relations, it is a stronger predictor of job performance in teamwork settings. They conclude from their analysis that "individuals who are more [...] secure and well adjusted are able to cooperate and work more effectively with others than those who scored lower on these characteristics" (Mount et al., 1998: 148). Barrick et al. (2001) find emotional stability to be a valid job performance predictor across different kinds of jobs (including teamwork). Hence, extant findings suggest that emotional stability is related to performance in teamwork.

Second, there is a relationship between emotional stability and the tendency of individuals to make use of new technologies (McElroy et al., 2007). Devaraj and co-authors (2008) find that emotional stable individuals are more likely to use and accept new technologies. These authors argue that emotional stable people are more likely to perceive a new technology as useful because they generally tend to react positively in both work and life situations. The perceived usesfulness of a new technology in turn affects the intention to use a new technology, and ultimately, the actual use of the new technology by the focal individual.

Third, emotional stability has also received attention in the literature as an antecedent of interpersonal interaction and relationships (e.g., Holland & Roisman, 2008). For instance, Wanberg and others argue that neurotic people do not engage as heavy in networking activities as their emotional stable peers because they are "more prone to self-consciousness, hostility, and complaining when seeking out others rather than using more constructive or problem-focused interactions" (2000: 493). Similarly, Asendorpf and Wilpers (1998) propose that neuroticism is negatively related to the formation of new relationships. However, neither Wanberg and co-authors (2000) nor Asendorpf and Wilpers (1998) have been able to demonstrate a relationship between emotional stability and social relationships empirically.

2.1.1.4 Introversion

The personality trait of introversion is sometimes also coined extraversion or surgency, thereby referring to the other end of the continuum of the trait. Introversion addresses interpersonal domains of personality (Wolff & Kim, 2012). It captures different facets, such as ambition, risk liking as well as activity (Judge et al., 1999). Most apparently, extraversion is related to sociability (Judge et al., 1999; Wanberg et al., 2000). Many scholars argue that extraverts are outgoing, talkative, gregarious, and like to interact with other people (Holland & Roisman, 2008; Mount et al., 1998; Wanberg et al., 2000). As Camgoz and colleagues put it: "Extraverted people tend to be enthusiastic, action-oriented individuals, who like to talk and assert themselves" (2011: 139). To the contrary, introverts are often described by terms such as shy, inhibited, submissive, and quiet (Brown et al., 2002; Holland & Roisman, 2008).

Beside sociability, scholars often stress that extraverts tend to be emotionally positive and optimistic (Holland & Roisman, 2008; Judge et al., 2002). For instance, McElroy and colleagues (2007) highlight that extraversion is associated with cheerfulness and Barrick and co-authors (2001) stress the excitement-seeking nature of extraverts. Furthermore, extraverts have a strong desire for success (Devaraj et al., 2008). In other words, extraversion encompasses also facets of amibition (Barrick & Mount, 1991). Related to that, the trait of extraversion is also associated with highly active, assertive, and energetic personalities (Camgoz et al., 2011; Wolff & Kim, 2012). For instance, Holland and Roisman (2008) describe extraverts as animated and lively. Moreover, extraverts tend to be less risk averse and adventurous (Mount et al., 1998) as well as high-spirited (Wanberg et al., 2000). I review the three literature streams of consequences of introversion (i.e., job performance, technology acceptance, and networking behavior) relevant in the context of this dissertation in the following.

First, a number of scholars have investigated the relationship between intro- and extraversion and job performance, often with a specific focus on teamwork or interpersonal tasks. For instance, Camgoz et al. (2011) argue that the risk-taking nature of extraverts is beneficial for their performance on the financial market. In line with their theoretical proposition, they find a positive relationship between the personality trait of extraversion and the financial performance of fund managers. Barrick and Mount (1991) find that extraversion is related to job performance in tasks during which interpersonal interaction is crucial. In their argument, they highlight the outgoing and talkative nature of extraverted personalities and find empirical support for the proposed relationship. Barrick and colleagues (2001) argue in a similar direction and find that extraversion is associated with success in teamwork.

Second, with regard to the usage and acceptance of new technologies, there is evidence that extraverts are more likely to adapt a new technology because they place enourmous value on their progressive image (Devaraj et al., 2008).

Third, numerous researchers have examined the relationship between extraversion and social relationships with other people. Wolff and Kim (2012) suggest a positive relationship between extraversion and networking behavior of individuals because of three underlying reasons. First, and most generally, extraversion can be described as a prerequisite for building social relationships because it describes a "tendency to approach social situations" (Wolff & Kim, 2012: 48). Second, because extraversion also entails assertiveness, extraverts do not hesitate to get in contact with other people. Finally, extraverts put effort in building social relationships as they enjoy attention from others. In line with their theoretical arguments, Wolff and Kim (2012) find a positive relationship between extraversion and networking behaviors. Similarly, Wanberg and her co-authors (2000) examine unemployed job seekers and their networking activities. They find that extraversion is a particular important antecedent of networking behavior as extraverts have a larger network to draw on. In addition, as discussed above, extraverts like to be in contact with other people.

2.1.1.5 Conscientiousness

There has been an intense debate in the literature concerning the essence of the conscientiousness factor of the "Big 5", reflected in a number of labels, e.g., conformity or "will to achieve" (Barrick & Mount, 1991). Today, researchers agree upon the fact that conscientiousness is of task-related nature (Wolff & Kim, 2012) and involves three different facets: (1) Achievement orientation, (2) dependability, and (3) orderliness (Judge et al., 1999). As Barrick et al. put it: "Conscientiousness is associated with dependability, achievement striving, and planfulness" (2001: 11). Similarly, Holland and Roisman (2008) describe conscientious people as focused, well organized, and responsible.

The first facet, achievement orientation, captures aspects such as hardworking and persistence (Judge et al., 1999). Related to that, Judge et al. (2002) argue that conscientiousness involves a tendency of individuals to working involvement. Similarly, Devaraj and others propose that "[c]onscientious personalities are intrinsically motivated to achieve, perform at a high level, and take actions to improve their job performance" (2008: 96) and Wanberg et al. (2000) relate conscientiousness to the term perseverance. The second facet, dependability, involves responsibility and carefulness (Judge et al., 1999). Barrick and Mount (1991) and Wanberg et al. (2000) describe conscientious people as careful, thorough and responsible. Finally, Judge and others (1999) suggest that conscientiousness contains aspects of orderliness – in other words conscientious people are planful and well organized. Extant researchers argue that conscientiousness is connected to efficient and systematic lines of action (Saucier, 1994) as well as to self-discipline (McElroy et al., 2007). Related to that, Camgoz et al. argue that conscientious people have "preferences for following rules and schedules" (2011: 139). Similar to the sections which discuss the other four factors of the FFM, I review studies that look at relationships between conscientiousness and job performance, technology acceptance, and networking behavior in the following.

First, with regard to job performance, there is evidence that consciencious fund managers are more successful as they often spend more effort on information acquisition and analysis than their peers (Camgoz et al., 2011). Similarly, Mount and Barrick (1998) study performance in jobs involving interpersonal interaction and find that conscientiousness is significantly related to achievement because it determines the ability to cooperate with others in teamwork and comparable settings. Barrick and Mount (1991) argue more generally that conscientiousness is beneficial for all kinds of jobs as it captures many characteristics that are helpful for task accomplishment, for instance planfulness and responsibility. Barrick et al. (2001) confirm this suggestion as they find that conscientiousness affects job performance positively across a number of occupations and performance measures. Judge et al. (1999) take a similar line and find that conscientiousness is positively related to intrinsic (i.e., job satisfaction) and extrinsic career success (i.e., monetary income or status).

Second, there are studies that examine the relationship between conscientiousness and acceptance of new technologies (McElroy et al., 2007). Devaraj et al. (2008) find that conscientiousness is positively related to the use of new technologies because of two reasons. First, because of the achievement-oriented nature of conscientious people, they will accept and adopt a new technology as long as they assess it as an efficient working tool. Second, conscientious people consider opinions of others with regard to new technologies more intensively. That is, if trustworthy people recommend the adoption of a new technology, conscientious individuals are more likely to follow their call as they have a strong desire to perform well.

Third, a number of studies have looked at conscientiousness as an antecedent of networking affairs. For instance, Wanberg et al. (2000) find a positive relationship between conscientiousness and networking activities of unemployed people. These scholars argue that conscientious unemployed people are highly motivated to find a new job because of their achievement orientation and, therefore, engage heavily in networking in order to accomplish that task. Interestingly, Wolff and Kim (2012) argue in another direction and suggest that conscientiousness may affect networking activities positively as well as negatively. For instance, the task-related nature of conscientious people may lead to pronounced networking activities to increase their job performance. On the other hand, as Wolff and Kim (2012) suggest, conscientiousness involves facets that might be negatively related to networking behavior. For instance, conscientious people might perceive networking as illegitimate as it could be interpreted as a means to bypass formal rules, thereby contradicting aspects of orderliness.

2.2 Motivations of innovation community members

A central characteristic of innovation communities is the voluntary engagement of community members in innovative processes (Roberts et al., 2006). Related to that, many scholars have investigated the motivations of individuals to take on innovation community activities. This section provides an overview of extant studies in this area and is structured in intrinsic and extrinsic motivations, a categorization which many scholars have used in the context of innovation communities (Borst, 2010; Janzik, 2010)[26]. Intrinsic motivation is at work if an individual "performs an activity for no apparent reward except the activity itself" (Deci, 1972: 113). In contrast, extrinsic motivation "refers to the performance of an activity because it leads to external rewards (e.g., status, approval, or passing grades)" (Deci, 1972: 113). As such, extrinsic motivation points to the consequences of an activity rather than the activity itself (Gagné & Deci, 2005; Grant, 2008). A number of scholars have investigated the interplay between intrinsic and extrinsic motivations (Deci, 1972; Gagné & Deci, 2005), an aspect that is not the focus of my research[27].

As qualitative evidence presented in part III suggests, extrinsic and intrinsic motivations of community members are important antecedents of innovation community social capital and affect the tendency of individuals to interact with many community members as well as to develop shared visions within the community. In this section, I review extant literature on intrinsic and extrinsic motivations of innovation community members. Based on these streams of research, I develop my hypotheses concerning the interplay between intrinsic and extrinsic motivations and innovation community social capital later on. Figure 20 summarizes my arguments.

[26] Of course, there are also other conceptualizations of motivations (e.g., Kehr, 2004).

[27] Many have argued that extrinsic motivation can be harmful for intrinsic motivation in open innovation settings (e.g., Füller, 2006), a phenomenon also referred to as the crowding out of intrinsic motivation (Frey & Oberholzer-Gee, 1997). Yet, there is also empirical evidence that extrinsic motivation can even affect intrinsic motivation positively (Roberts et al., 2006). I argue that my model does not contradict extant research on intrinsic and extrinsic motivations. Many scholars in the field of innovation communities have argued that extrinsic and intrinsic motivation exist simultaneously (see Borst (2010) for a review). Furthermore, I allow for correlation between intrinsic and extrinsic motivation in the SEM presented in this part, a common procedure in covariance-based models (Weiber & Mühlhaus, 2010).

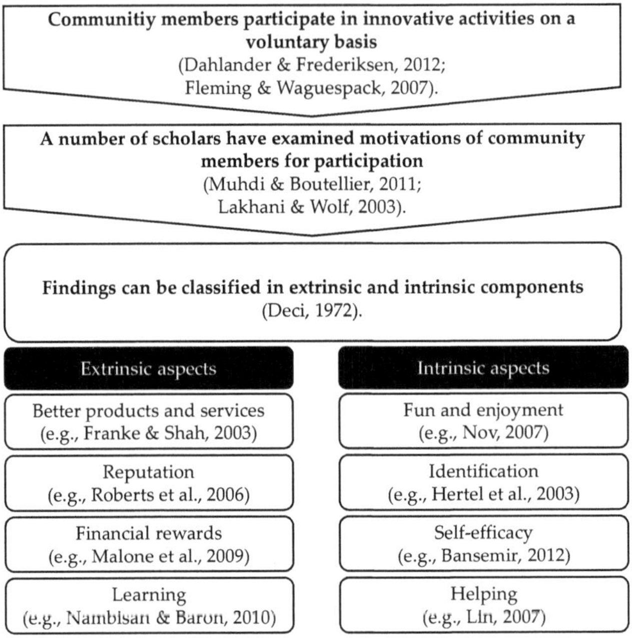

Figure 20: Intrinsic and extrinsic motivations of innovation community members

2.2.1 Extrinsic motivation

Extrinsic motivation is related to consequences of the focal task as opposed to the task itself. As such, extrinsic motivation is present "when the motives are located outside of an actual activity" (Janzik, 2010: 252). Such motivations are especially important because extrinsic motivations constitute levers that can be influenced by managers of innovation communities (Leimeister, Huber, Bretschneider, & Krcmar, 2009). My categorization of extrinsic motivation is similar to the arguments of other scholars (Janzik, 2010; Muhdi & Boutellier, 2011) and distinguishes between four categories: (1) Usage of better products and services, (2) reputation and career perspectives, (3) financial payment, and (4) learning and gathering knowledge. These four categories are discussed in the following.

2.2.1.1 Usage of better products and services

One important extrinsic motivation of community members is to benefit from better products or services (Lerner & Tirole, 2002). For instance, more than 40% of members of sports communities highly agreed upon the statement that they want to use the products developed by the community themselves (Franke & Shah, 2003). As another example, Hertel, Niedner, and Herrmann (2003) find that members of the Linux kernel community were often motivated by the pragmatic goal to program software which they are going to use by themselves. Similarly, Lakhani and Wolf (2005) study OSS communities and find that a majority of community members (more specifically 58%) contribute because they want to benefit from the improved code. Obviously, this motivation is not present in all kinds of innovation communities. For example, Füller (2006) argues that users who are members of innovation communities supported by big corporations such as Procter & Gamble will hardly experience such a motivation as the likelihood that their idea is going to be implemented is close to zero.

2.2.1.2 Reputation and career perspectives

Members also engage in communities because they desire recognition from others for their work (Lerner & Tirole, 2002; Malone, Laubacher, & Dellarocas, 2009). In some communities, for instance in communities in the context of sports, these reputation related motives do not seem to play a pivotal role (Franke & Shah, 2003). In other settings, however, motivations related to reputation are a pivotal driver to engage in community activities. For instance, Greer and Lei (2012) review extant literature and argue that consumers are often motivated to innovate because they aim for recognition of peers. Lakhani and Wolf (2005) identify a significant relationship between caring about reputation in OSS communities and time spent for community work. Likewise, Roberts and colleagues find that status motivations are positively related to participation in OSS projects because OSS contributors see community participation as a way to "signaling their programming abilities and competencies" (2006: 989).

However, competition for reputation may affect community work both positively as well as negatively. For instance, Frey and Lüthje (2011) argue that competition for recognition may lead to increased efforts of community members to outperform each other as well as to decreased interaction within the community. More specifically, community members may not share their ideas with their peers as they fear disadvantages. Frey and Lüthje (2011) find that competition for reputation in an end-user innovation community leads in those cases to higher quality of contributions to community work when competition for recognition is complemented by trust between community members.

The opportunity to increase ones reputation in the context of innovation communities is often related to career perspectives of community members. Recently, Bitzer and Geishecker (2010) looked at workplace-related motives of members of OSS communities. Interestingly, these scholars find that professional activities in OSS projects and voluntary engagement in similar settings during leisure time are highly correlated. Another study of Lakhani and Wolf (2005) argues in a similar direction and finds that 55% of OSS community members contribute to the community during working hours. Likewise, Hertel and his colleagues (2003) find that OSS programmers are often motivated by career perspectives. Related to that, Leimeister et al. (2009) analyze a community-based innovation contest supported by the German Software company SAP and find that appreciation by SAP is the most important motivation of community members. Bitzer and Geishecker conclude that there is "some support to the idea that career motives play a role in voluntary OSS activities" (2010: 171).

2.2.1.3 Financial rewards

Financial rewards are often discussed as extrinsic motivations in innovation community settings (Greer & Lei, 2012; Malone et al., 2009). With regard to financial payments, Lakhani and Wolf (2005) find that getting payed for community work affects the hours spent for community activities positively and significantly.

The extent to which financial rewards are important motivations differs between innovation communities. For instance, in their study in the field of innovation communities of sports enthusiasts, Franke and Shah (2003) find that financial compensation was only a minor motivation for community members to participate. Likewise, financial rewards only play a role of minor importance in consumer innovation communities (Füller, 2006). However, Füller (2006) also finds that financial rewards are important for those community members who spend a lot of time and effort for community activities.

In those settings where financial rewards are important motivators, researchers have elaborated on the detailed design of reward structures and incentives. For instance, Bothner, Podolny, and Smith (2011) investigate reward structures in communities engaged in science and research, an activity closely related to innovation. Bothner and co-authors (2011) discuss whether it is preferable to reward high-status scientists with greater payoffs than their low-status community peers. While this procedure is obviously not optimal with regard to social welfare, it is advantageous in specific contexts as it leads to huge incentives to become a high-status community member. More precisely, a concentration of rewards to high-status community members is beneficial in communities "comprised either of a handful of latent superstars" or of individuals "deeming [...] [their] work process inherently valuable" (Bothner et al., 2011: 441).

In many cases, companies use the concept of an innovation contest to encourage communities to engage in innovative activities (Bullinger et al., 2010; Hutter, Hautz, Füller, Mueller, & Matzler, 2011). In this context, Boudreau, Lacetera, and Lakhani (2011) take a look at the top-coder OSS community and the contests organized within this community. According to these scholars, the number of participants within a community has direct implications for the motivations induced from financial rewards. More specifically, two countervailing forces are present. First, a greater number of community members increases rivalry and, hence, the likelihood to receive rewards decreases and community members are likely to reduce efforts for community work. To the contrary, if more members engage in community activities, the propability of an extremely valuable outcome increases. Which of these two forces is more pervasive is contingent upon situational factors. More precisely, "the effort-reducing effect of greater rivalry dominates for less uncertain problems, whereas the effect on the extreme value prevails for more uncertain problems" (Boudreau et al., 2011: 843).

2.2.1.4 Learning

Another important extrinsic motivation in innovation communities is gathering of knowledge from other community members (Leimeister et al., 2009), a motivation some scholars have referred to as self-actualization (Greer & Lei, 2012). For instance, Franke and Shah (2003) find that innovation community members often share their ideas with other members because they expect that others will share their knowledge with them in the future. Related to that, members of consumer innovation communities reported that they are motivated to engage in community activities because they aim to get access to knowledge of other community members, thereby increasing their own skills and capabilities (Füller, 2006). Muhdi and Boutellier (2011) find that gaining knowledge from other community members is an important motivation for members of internal as well as external innovation communities. Nambisan and Baron (2010) find that expectations of expertise enhancement are positively related to knowledge contribution in community activities. In their often cited article on the topic, Lakhani and von Hippel (2003) provide empirical evidence from the Apache web server software community that gaining knowledge from other community members constitutes a major motivation of members of the innovation community. Within the Apache community, "98%, on average, of the time spent at the help website by an information provider is devoted to reading posted questions, and only 2% to providing answers" (Lakhani & von Hippel, 2003: 924). In the argument of Lakhani and von Hippel (2003), this explains an important motivation for contributing to the help website because contributors gather knowledge from other community members while they read their contributions. In sum, while individuals contribute to innovation communities, they exercise their own knowledge and skills (Nov, 2007). As a result, they learn about the issues discussed in the community they engage in. This learning process, in turn, motivates community members to engage in community activities.

Often, an important underlying assumption of the extrinsic motivation of learning in innovation communities is that direct and/or indirect reciprocity is present within communities (Faraj & Johnson, 2011). In other words, if a community member shares knowledge with another community member, the latter is likely to share his or her knowledge, too. Related to that, Kim, Lee, Zmund, and Bock (2005) look at knowledge exchange processes among employees within a company and find that the relationship between anticipated reciprocity and an individual's intention to share knowledge is mediated by the attitude of the focal individuum towards knowledge sharing. Lin (2007) finds that expected reciprocal benefits are positively related to knowledge sharing intentions as well as to attitudes toward knowledge sharing.

2.2.2 Intrinsic motivation

Intrinsic motivation arises from "a natural internal desire or from feelings of competence, satisfaction, or fulfillment" (Leimeister et al., 2009: 208). As a result, intrinsically motivated community members contribute to the community "for its own sake rather than as a means to an end" (Roberts et al., 2006: 985). Intrinsic motivation is associated with creativity (Grant & Berry, 2011), an ability important in settings such as innovation communities. Thus, many scholars have argued that intrinsic motivation is particularly important in these environments (Franke & Shah, 2003; Füller, 2006). Janzik (2010) reviews extant literature on motivations of innovation community members. In his summary of extant publications, he highlights that a great majority of extant studies finds that intrinsic facets represent very strong motivations to engage in innovative activities. Related to that, intrinsic motivations are also present in community activities where extrinsic rewards are the dominant driver of community participation. For instance, Boudreau, Lacetera, and Lakhani (2011) examine an OSS community which regularly rewards successful community members extrinsically within contests. Even in such environment, motivations behind cash-prizes, such as fun and pleasure to compete with other community members, are important. Intrinsic motivations in community settings can be categorized in four facets (see Janzik (2010) and Nambisan and Baron (2009) for similar arguments): (1) fun and enjoyment, (2) identification with the community, (3) increased knowledge self-efficacy, and (4) positive emotions related to helping others. These facets are discussed in the following sections.

2.2.2.1 Fun and enjoyment

One important intrinsic motivation to join innovation communities is enjoyment, fun, and interest in the innovative task of the community (Malone et al., 2009). As Greer and Lei put it: "The actual creative process, including the interaction experience, the challenge of the problem and the potential for cognitive stimulation, also has reward implications" (2012: 71). In a similar vein, Nov (2007) investigates motivations of contributors to Wikipedia, the popular and often explored user-created encyclopedia, and finds that having fun is the most important motivation of Wikipedians. According to extant publications, fun and enjoyment of community members stems from three different sources.

First, community members are motivated to engage in community activities because they enjoy the *exchange with other community members*. Franke and Shah (2003), for instance, find that most community members agree that it is fun to them to create something together with other community members. In addition, nearly a third of all community members who took part in the survey highly agreed upon the statement that they like to be an expert-adviser for other community members. Likewise, Nambisan and Baron (2009) identify hedonic aspects as significant motivations of customers to engage in innovation community activities. According to these scholars, participation in innovation communities is associated with fun and pleasure because community members enjoy the exchange with other community members.

Second, innovation community members are motivated because they are *highly interested in the task that is accomplished within the community* and, as a result, have fun while performing community activities. For example, Lerner and Tirole (2002) investigate motivations of OSS communities and argue that community members compare the tasks of the community work to other tasks they could do at the same time for their employers. In the mind of these researchers, "the programmer compares how enjoyable the mission set by the employer and the open source alternative are. A 'cool' open source project may be more fun than a routine task" (Lerner & Tirole, 2002: 213). Similarly, Füller studies consumer innovation communities and finds that as "consumers' engagement is voluntary, interest in the task is the most necessary prerequisite" (2006: 642).

Third, the *creative nature of community work* is often perceived as intellectually stimulating and, hence, community members have fun while engaging in such activities. In this context, Lakhani and Wolf (2005) sent a survey to OSS developers and find that 45% of open source community members contribute because they experience community work as intellectually stimulating. Similarly, Nambisan and Baron (2009) argue that the problem-solving nature of community activities is perceived as pleasurable because it represents an intellectual challenge for community members that is associated with fun.

2.2.2.2 Identification with the community

Similarly, other scholars propose that identification with the community is a strong motivation to contribute to innovation communities (Lakhani & Wolf, 2005). For instance, Hertel, Niedner, and Herrmann (2003) find that most important motivations of members of an OSS community are related to identification as a user or a programmer of that OSS.

Related to that, other scholars argue that community members often share their knowledge with others in the community because they feel a moral obligation or responsibility to do so (Nambisan, 2002). According to McLure Wasko and Faraj (2000), community members exchange knowledge intensively because they conceptualize knowledge as a public good. As a result, community members are motivated to contribute their knowledge and ideas to community work as they think that giving knowledge back to the community is the right thing to do. I argue that this aspect is closely related to identification with the community because it describes how community members perceive themselves as part of a larger group.

Related to that, Bagozzi and Dholakia (2006) study motivations of a software developer community and find that we-intentions to participate in the community are an important motivation to interact within the community, to visit related websites, or to spend money for community activities. According to the authors, "we-intentions are formulated when individuals think of themselves as 'us' or 'we' and remain jointly ready to act in activities to accomplish the group's collective goals" (Bagozzi & Dholakia, 2006: 1103). As such, the construct of we-intentions to participate constitutes an intrinsic motivation to participate in innovation communities.

Another facet that is related to identification with the community is passion for the community and the work that is accomplished in community settings. For instance, Faraj et al. (2011) identify passion for community goals as an important motivation to share knowledge with other community members in online settings. Passion for community goals leads often to the impression of community members to add to something that is of high importance and, hence, motivates community members to put effort in innovative activities (Malone et al., 2009). Devoted enthusiasm towards community activities leads to entrepreneurial behavior and to a greater amount of time spent within the community. However, Faraj and his co-authors (2011) also discuss negative consequences of passion within communities. For instance, passion often is associated with reduced willingness to compromise and, thus, leads also to a greater potential of conflict among community members.

2.2.2.3 Increased knowledge self-efficacy

Another intrinsic motivation of individuals to share their knowledge with others in innovation community settings is gaining of knowledge self-efficacy (Lin, 2007). Knoweldge self-efficacy refers to an individual's self-consciousness with regard to the usefulness of his or her knowledge for others. Gist and Mitchell define self-efficacy as "a person's estimate of his or her capacity to orchestrate performance on a specific task" (1992: 183). An individual who contributes to innovation community activities receives the feeling of being valuable for the community in exchange for his or her contribution, thereby enhancing knowledge self-efficacy (Lin, 2007). Related to that, Bansemir et al. (2012) find that technological induced self-efficacy in intra-organizational innovation communities is related to engagement in knowledge sharing. In an experimental pretest-posttest study, they find that those individuals who read a message that tells them that they can contribute successfully to the community engage more heavily in community activities.

2.2.2.4 Positive emotions related to helping

Furthermore, engaging in community activities includes the feeling of helping (McLure Wasko & Faraj, 2005). As the feeling of being helpful is generally related to positive emotions, this aspect can also be considered as an intrinsic motivation of innovation community members. Lin (2007) finds that people who perceive helping as something positive have more positive attitudes towards knowledge sharing as well as higher knowledge sharing intentions. Grant takes a more nuanced view and conceptualizes helping others as a prosocial motivation, which he defines as "the desire to expend effort to benefit other people" (2008: 49). The literature on motivations in innovation community settings, however, often assigns aspects related to helping others to intrinsic motivations.

3 Model development

In this chapter, I present my hypothesized model. The chapter is organized as follows. First, I present my hypotheses concerning the relationships between different dimensions of social capital. Second, I present my hypotheses for the relationships between the "Big 5" personality traits and structural as well as cognitive community social capital. Finally, I present hypotheses on the relationships between motivations of community members and structural as well as cognitive social capital in corporate innovation community settings. The hypothized model is depicted in Figure 21. The conceptualization of social capital in the model is similar to Tsai and Ghoshal (1998). More specifically, the model treats network size (i.e., structural social capital) as well as shared visions and languages (i.e., cognitive social capital) as antecedents of tie strength as well as trust (i.e., relational social capital).

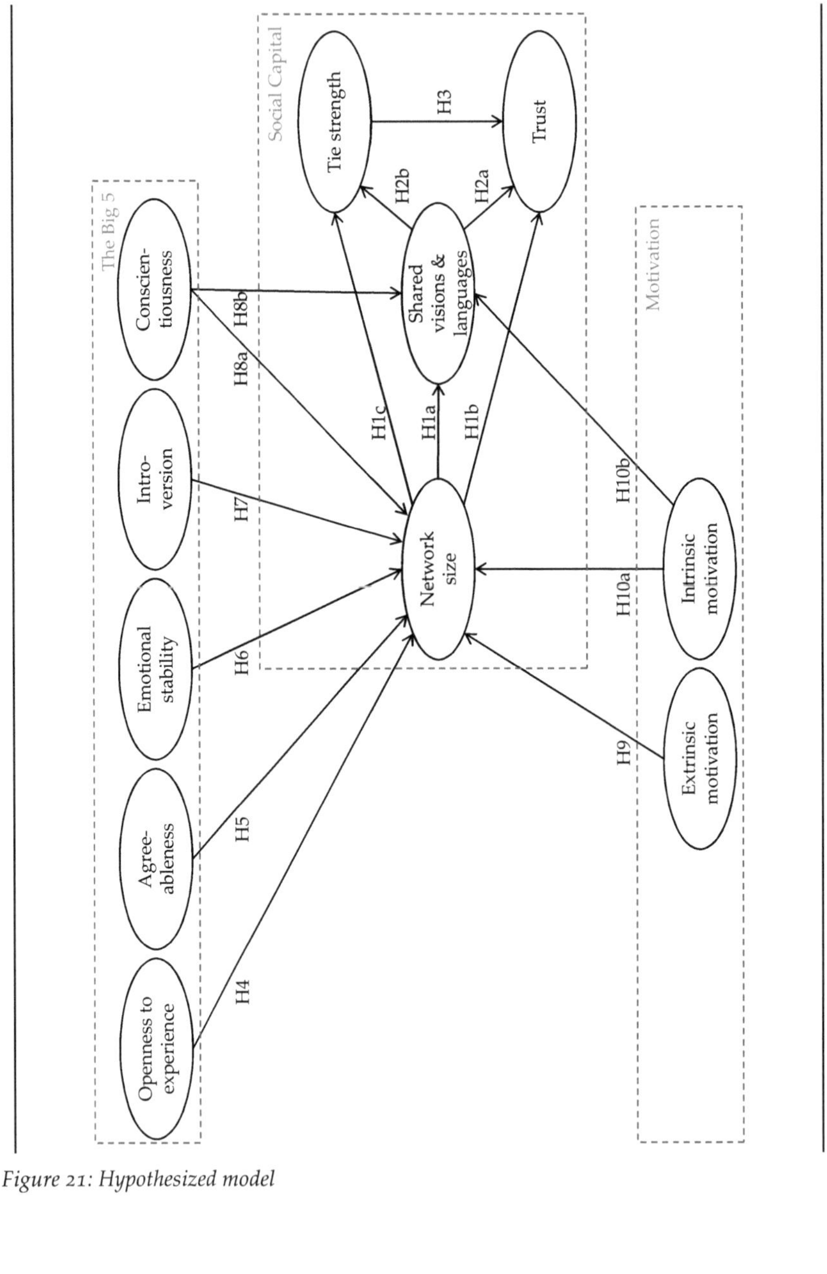

Figure 21: Hypothesized model

3.1 Self-reinforcing relationships between structural, cognitive, and relational social capital in corporate innovation communities

Section 3.1 presents my hypotheses with regard to the relationships between different dimensions of social capital[28]. The conceptualization of social capital used relies on Nahapiet and Ghoshal (1998) who distinguish between structural, relational, and cognitive dimensions of social capital. The relationships between the three dimensions of social capital are closely aligned with the study of Tsai and Ghoshal (1998), whose model is extended in this section. In doing so, I follow the call of Inkpen and Tsang (2005) for more research on interaction effects between different dimensions of social capital.

More precisely, the model presented conceptualizes social capital as follows. The structural dimension of social capital is conceptualized as the network size of community members, an often used measure in the literature (Carpenter et al., 2012). The cognitive dimension in my model captures, as proposed by Nahapiet and Ghoshal (1998), aspects related to shared visions and languages. With respect to the relational dimension, I take a more nuanced view than Tsai and Ghoshal (1998). In contrast to these scholars, the model presented distinguishes between two different constructs of relational social capital, i.e., trust as well as tie strength (Maurer et al., 2011). It follows the arguments of Levin and Cross (2004) who have shown that these two constructs represent distinct phenomena. I use the terms network size (for structural social capital), shared visions and languages (for cognitive social capital), as well as tie strength and trust (for relational social capital) in the following in order to develop the hypothesized relationships. Figure 22 summarizes the conceptualization of social capital in this part of the dissertation. The next section presents my hypotheses.

[28] For a detailed review of extant literature on social capital, see part II.

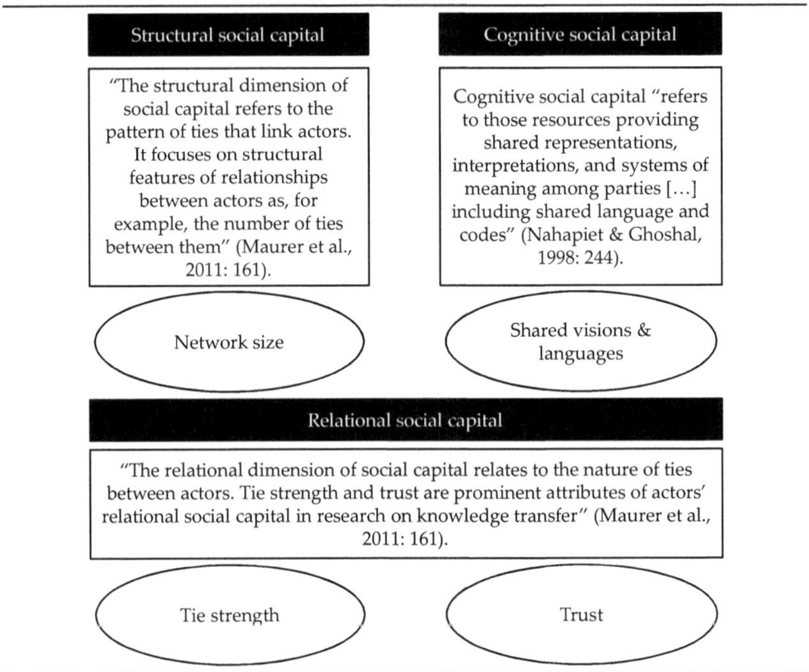

Figure 22: Conceptualization of social capital

3.1.1 Network size and shared visions, tie strength as well as trust

I propose that network size of community members is positively related to a shared vision and language with other community members. Community members build shared visions and languages while they work together (Maurer & Ebers, 2006; McLure Wasko & Faraj, 2005; Nahapiet & Ghoshal, 1998). In the words of Tsai and Ghoshal: "The association between the structural and the cognitive dimensions of social capital relies on the premise that social interaction plays a fundamental role both in shaping a common set of goals and values and in the sharing of those goals and values among an organization's members" (1998: 466). Coleman (1988) argues in a similar direction as he suggests that networks characterized by multiple ties between actors are likely to result in shared norms within the network. Thus, the bigger the interaction network of a community member, the more likely the focal individual will develop shared visions and languages within the community. In addition, those at the center of the innovation community network will be able to influence shared visions of the community to a significant extent. For instance, the study of Krackhardt (1992) demonstrates how those that possess a central network position (i.e., a high number of ties) in an intraorganizational network influence shared attitudes towards an unionization attempt. Based on these arguments, I propose

Hypothesis 1a: Network size of a community member is positively related to shared visions and languages of the focal community member with other members.

In addition, I suggest that network size affects trust among community members. Two related theoretical lines of argument underline this hypothesized relationship. First, similar to the arguments with regard to Hypothesis 1a, trust develops while people interact with one another (Nahapiet & Ghoshal, 1998; Tsai & Ghoshal, 1998). As Blatt puts it: "[P]eople working in a corporation develop trust toward one another as they accumulate detailed information about each other over time" (2009: 535). Second, network size determines an individual's ability to diffuse information in the network (Rogers, 1995). In consequence, individuals with numerous contacts are able to establish a reputation of a trusted person more easily as these individuals can draw on many advocates within the network (Wong & Boh, 2010). Hence, community member's network size is positively related to the development of trust within the community as community members who interact with many others have a multitude of opportunities to develop trustful relationships. In addition, members at the center of the community network benefit from the information their contacts diffuse within the community with respect to their trustworthiness. In more formal terms

Hypothesis 1b: Network size of a community member is positively related to trusted relationships of the focal community member with other members.

Furthermore, I suggest that network size is positively related to tie strength. Tie strength describes the closure and frequency of interaction with other community members (Hansen, 1999). By definition, strong ties develop based on initial knowledge and resource exchange and some form of interaction is a prerequisite for the development of strong ties (Jack, 2005). Recently, Levin, Walter, and Murnighan introduced the concept of dormant ties, i.e., "a relationship between two individuals who have not communicated with each other for a long time" (2011: 923) and find that these dormant ties are beneficial because they constitute opportunities to reconnect to individuals. In line with my arguments, Levin et al. (2011) suggest that existing ties constitute opportunities to deepen relationships with individuals. Similarly, Gargiulo and Benassi (2000) suggest that numerous ties within a network enhance cooperation. Hence, those individuals in the community that interact with many other community members and possess a big network are likely to develop stronger ties to other community members as their existing network provides opportunities to develop stronger relationships. That is, strong ties can only be developed in an existing network. I propose

Hypothesis 1c: Network size of a community member is positively related to the strength of ties of the focal community member to other members.

3.1.2 Shared visions and trust as well as tie strength

I propose that a shared vision and language affects trust within an individual's network positively. A number of scholars have argued for a beneficial influence of cognitive social capital on trust within a network (e.g.,Tsai & Ghoshal, 1998). In more detail, shared visions are associated with appropriate behavior (Hughes & Perrons, 2011), an aspect which is associated with trust among community members. In this context, Nahapiet and Ghoshal (1998) talk about an accumulation of cognitive and relational social capital. Similarly, Oh et al. (2004) propose that trust among network members develops in environments where members agree upon shared visions. In innovation communities, a shared vision on community activities provides an environment where trust is more likely to be present as community members are united by a shared goal towards each community member works. I propose based on these theoretical arguments

Hypothesis 2a: Shared visions and languages of a community member with other members are positively related to trusted relationships of the focal community member with other members.

Moreover, cognitive social capital is associated with stronger ties to other community members. Cognitive social capital captures both a shared vision and language with other community members (Nahapiet & Ghoshal, 1998). Both of these aspects lead to increased frequency and closeness of interaction (i.e., tie strength) with other community members. As Lechner et al. put it: A "shared vision creates a hospitable social context by [...] improving the efficiency of communication and tacit knowledge transfer" (2010: 883). With respect to shared languages, Bolino et al. state: "Where shared language and shared narratives both exist, employees can more easily discuss problems, transfer ideas, share knowledge, and offer more effective assistance to one another" (2002: 511). Hence, a shared vision implies that community members pursue a common goal with their network ties, and therefore, cooperate more frequently and closely with other community members. In addition, a shared language facilitates exchange, and therefore, those who share a language interact more frequently and more closely. In formal terms

Hypothesis 2b: Shared visions and languages of a community member with other members are positively related to the strength of ties of the focal community member to other members.

3.1.3 Tie strength and trust

I propose that tie strength (i.e., frequency and closeness of interaction) is positively related to the development of trust. Many scholars have argued that trust developes while people interact with one another (Blatt, 2009). In essence, people use past interactions with others to evaluate the trustworthiness of their peers. As Blatt puts it: "The longer the track record of interaction, the more opportunities people have to demonstrate their investments and reciprocity in the relationship, all of which foster trust" (2009: 535). Related to that, Fang (2011) argues that intense collaboration in innovative strategic alliances is beneficial as it is associated with trust. The study of Levin and Cross (2004) furthermore supports a relationship between tie strength and trust. These scholars find that the relationship between tie strength and receipt of useful knowledge is mediated by both competence-based and benevolence-based trust. Hence, close and frequent interaction among community members in innovative settings is beneficial for the emergence of trusted relationships between community members. I propose

Hypothesis 3: The strength of ties of a community member to other members is positively related to trusted relationships of the focal community member with other members.

3.2 The "Big 5" personality traits as antecedents of corporate innovation community social capital

This section presents my hypotheses on the relationship between openness to experience, agreeableness, emotional stability, introversion, and conscientiousness (i.e., the "Big 5" personality traits) and innovation community social capital. My arguments are supported by Anderson (2008) who stresses the importance of personality for social interaction and relationships.

3.2.1 Openness to experience and innovation community social capital

I argue that openness to experience of community members is positively related to community members' network size (i.e., the number of network ties a member possesses within the community). This hypothesis is based on four arguments. First, because open individuals are often described as creative and innovative (Saucier, 1994), they enjoy the setting of innovation communities as well as exchanging and networking activities within these. Thus, their network size is high. Second, and related to that, the findings of the literature on networking behavior indicate that building ties to people in order to gain access to relevant knowledge and ideas is a behavioral pattern often used by open individuals (e.g., Wolff & Kim, 2012). Third, extant scholars argue that individuals who score high on openness to experience are good at learning new things (Barrick & Mount, 1991). The literature on communities of practice argues that learning and innovating are closely related (Brown & Duguid, 1991). Thus, open individuals are good at exchanging knowledge and ideas with known and unknown people and therefore, create a high number of social ties in innovation communities. Fourth, openness to experience is related to the use of new technologies (Devaraj et al., 2008). The use of new technologies such as community platforms allows open people to create new ties as well as to sustain existing ties online. Consequently, I propose

Hypothesis 4: Openness to experience of a community member is positively related to the focal community member's network size.

3.2.2 Agreeableness and innovation community social capital

I furthermore argue that agreeableness is positively related to network size. My hypothesis is based on two underlying mechanisms. First, the tendency of agreeable personalities to invest in social relationships even at an expense of productivity fits perfectly to the setting of corporate innovation communities. The literature has emphasized that a too strong focus on productivity is harmful for innovation because at the outset, it is not possible to determine whether a good idea will be useful in the end (van de Ven, 1986). As such, agreeable people will be successful in establishing network ties in innovation communities because their focus on exchange perfectly fits to community activities. Second, the tendency of agreeable people to make use of new technologies contributes to the formation of ties within innovation communities. This argument is particularly tempting because Devaraj and colleagues (2008) stress that agreeableness leads to more intensive use of technologies if the focal technology enables cooperation and exchange with others. As online innovation community platforms essentially are there to support exchange and cooperation between members, agreeable individuals will make more use of these technologies and this, in turn, helps them to build and sustain network ties within the community. In more formal terms, I propose

Hypotheses 5: Agreeableness of a community member is positively related to the focal community member's network size.

3.2.3 Emotional stability and innovation community social capital

I suggest that emotional stability affects the network size of community members positively. In other words, emotional stable people are likely to own a more central network position than their neurotic peers. Two underlying theoretical lines of argument explain this relationship. First, extant research has shown that emotional stable people perform good with regard to tasks that ask for intense personal interaction such as teamwork (Barrick et al., 2001; Mount et al., 1998). The ability to interact and communicate with other people is beneficial in innovation communities because these behavioral patterns describe important activities in innovation communities - to exchange knowledge and ideas with known and unknown community members. My arguments are somewhat contradictive to the findings of extant empirical studies in the literature on the relationship between emotional stability and networking activities. For instance, Wanberg et al. (2000) have investigated the relationship between emotional stability and networking activities of unemployed people and found no significant relationship. Similarly, Asendorpf and Wilpers (1998) look at networking behavior of university students and Wolff and Kim (2012) look at informal networking activities of employees – both of them find no significant influence of emotional stability on networking behavior. However, I argue that these findings are not transferable to the context of innovation community activities. In essence, the empirical settings of the studies described above conceptualize networking behavior as an informal activity, either implicitly or explicitly. In contrast, networking in innovation communities is a major part of community work and because of that, the distinction between formal and informal communication is not applicable in community settings.

Second, emotional stability affects structural social capital as emotional stable people make more intense use of new technologies (Devaraj et al., 2008). This is beneficial for the number of ties within communities because community activities take place both in offline as well as in online settings. Those who use online platforms more heavily have better opportunities to build social ties, especially to people who are geographically distant. In formal terms

Hypothesis 6: Emotional stability of a community member is positively related to the focal community member's network size.

3.2.4 Introversion and innovation community social capital

I propose that introversion is negatively related to network size. My arguments are based on the qualitative research findings presented in part III as well as on extant studies of numerous scholars who examine intro- and extraversion as an antecedent of related concepts to social capital, e.g., networking activities (Wolff & Kim, 2012) and performance in teamwork (Barrick et al., 2001). Introversion affects network size mainly because of three underlying theoretical forces. First, introverts are shy and avoid conversations with others (Mount et al., 1998; Wanberg et al., 2000). Thus, they will also be less likely to get in contact with other community members. Second, the passive nature of introverts affects network size negatively as extraverted people will pursue their innovative task within the community more actively, thereby exchanging ideas with many people. Finally, introverts will have a smaller network within innovation communities as they use new technologies less intensively (Devaraj et al., 2008). As many community activities take place based on IT, usage of new technologies will also affect the number of ties within the community. In more formal terms, I propose

Hypothesis 7: Introversion of a community member is negatively related to the focal community member's network size.

3.2.5 Conscientiousness and innovation community social capital

I propose that conscientiousness is positively related to both network size of the focal community member as well as shared visions and languages with other community members. Conscientious people form more ties in community settings because of several reasons. First, because conscientious people are achievement-oriented (Judge et al., 1999), they try to acquire knowledge and ideas from a large number of other individuals. Second, conscientiousness is related to carefulness, in other words a tendency of individuals to perceive a high responsibility to accomplish a certain task (Wanberg et al., 2000). This feeling of responsibility also affects network size positively because it also leads to a tendency to acquire knowledge and other resources from many individuals in order to perform well with regard to community activities. Finally, related to the arguments of Devaraj et al. (2008), conscientious people build more social ties in innovation community settings as they rely more heavily on IT that supports knowledge and idea exchange. In formal terms, I propose[29]

Hypotheses 8a: Conscientiousness of a community member is positively related to the focal community member's network size.

Furthermore, I argue that conscientiousness is positively related to shared visions and languages with other community members. The responsible and careful nature of conscientious people is beneficial for the development of shared visions. Furthermore, conscientiousness is associated with high identification and feelings of responsibility for the result of community activities (Judge et al., 1999; Wanberg et al., 2000). These aspects will also lead to a feeling of a shared vision that links a conscientious individual to its network ties. In addition to a shared vision, social capital scholars often argue that the cognitive dimension of social capital also entails facets of shared vocabulary and language (e.g., Packalen, 2007). The high identification with community activites induced by the responsible nature of conscientious individuals is likely to lead also to the development of a unifying language. I propose

Hypotheses 8b: Conscientiousness of a community member is positively related to shared visions and languages of the focal community member with other members.

3.3 Extrinsic and intrinsic motivations as antecedents of corporate innovation community social capital

This section presents my hypotheses concerning extrinsic and intrinsic motivations and innovation community social capital. In more detail, extrinsic motivation is proposed to affect network size of a focal individual. In addition, intrinsic motivation is hypothesized to be positively related to network size as well as shared visions and languages with other community members. My hypotheses and the underlying theoretical arguments are presented below.

---

[29] It is interesting to note that the arguments of Wolff and Kim (2012) for a negative influence of conscientiousness on networking activities are not relevant in an innovation community setting. In innovation communities, there are no formal rules present that inhibit activities to build social ties.

3.3.1 Extrinsic motivation and innovation community social capital

I suggest that extrinsic motivation in innovation community settings is positively related to network size because of a number of reasons. First, extrinsic motivation involves the desire to benefit from a better reputation as well as better career perspectives in the future (Lerner & Tirole, 2002). An individual who participates in community activities because he or she wants to benefit from better reputation is likely to contact more peers within the community. In essence, it is hard to improve one's reputation without getting in touch with others. In addition, extrinsic motivation captures a strong interest in monetary or non-monetary rewards (Greer & Lei, 2012). Individuals who want to win a prize within a community, for instance in the context of an innovation contest, are likely to try to receive as much knowledge as possible from other community members as they want to receive the (financial) rewards. Hence, extrinsically motivated individuals try to contact many community members in order to benefit from their knowledge and ideas. Related to that, extrinsic motivation captures a person's desire to learn from his or her peers within the community (e.g., Lakhani & von Hippel, 2003). As a result, the motivation to learn from others is also related to the tendency of getting in contact with many others in the community. In formal terms

Hypothesis 9: Extrinsic motivation of a community member is positively related to the focal community member's network size.

3.3.2 Intrinsic motivation and innovation community social capital

I propose that intrinsic motivation is positively related to network size, mainly because of two underlying theoretical lines of thought. First, intrinsic motivation in innovation communities captures an individual's predisposition to enjoy the exchange with other people (Franke & Shah, 2003). Such a tendency leads to more ties in community settings because intrinsically motivated individuals actively search for additional ties in order to exchange knowledge. Second, intrinsic motivation includes a person's tendency to enjoy helping others (Lin, 2007). An individual who engages heavily in helping behavior in community settings is likely to get in contact with a large number of other community members. I propose

Hypothesis 10a: Intrinsic motivation of a community member is positively related to the focal community member's network size.

Moreover, I propose that intrinsic motivation is positively related to a shared vision and language with other community members. In essence, intrinsic motivation describes motivations of a focal individual directly related to community activities as opposed to an outcome of community activities that is detached from the actual task of community work (Deci, 1972). As such, intrinsic motivation implies that an individual perceives the outcome of a community task as his or her goal, and this goal represents a shared vision with other community members. Moreover, intrinsically motivated individuals identify themselves stronger with the community, a tendency that some scholars have described by the term "we-intentions" (Bagozzi & Dholakia, 2006). Obviously, such "we-intentions" are related to a shared vision with other community members. I propose

Hypothesis 10b: Intrinsic motivation of a community member is positively related to shared visions and languages of the focal community member with other members.

4 Empirical field

In this chapter, I describe the empirical field data was collected at in order to test the model presented. I organize my arguments in two sections. In section 4.1, I describe the empirical setting of my study, the WIN Innovation Community, and discuss how the WIN Innovation Community conforms to the definition of corporate innovation communities presented in part II. In section 4.2, I describe the data collection procedure.

4.1 WIN Innovation Community

The WIN Innovation Community is a corporate innovation community supported by the Institute of Information Systems at FAU. At the start of the winterterm, first year students at the School of Business and Economics are invited to engage in community activities. Data was collected in the winterterm 2011/2012, when the community consisted of 1,746 members. The Institute of Information Systems announces different tasks to the community, which are solved in the context of innovation contests[30]. In the winterterm 2011/2012, the students were set with the task to develop an innovative concept, which relies on new technological possibilities in the context of tablet PCs and helps to realize potentials in one of three different fields: (1) Healthy nutrition, (2) e-learning solutions, and (3) in the context of the federal volunteer service. In the following, I discuss how the WIN Innovation Community conforms to the definition of a corporate innovation community presented in part II. I take all five characteristics of innovation communities into consideration.

[30] There are numerous examples where the instruments of innovation communities and innovation contests are combined in practice. For instance, the open source community Topcoder uses contests among its members in order to solve specific problems (e.g., Boudreau et al., 2011). The WIN Innovation Community is also an example of a community within which innovation contests are organized. Other publications on the WIN Innovation Community have focused their research efforts on these contests (e.g., Adamczyk, Haller, Bullinger, & Möslein, 2011). The ambivalence of the WIN Innovation Community is reflected in the work of Bullinger et al. (2010), who describe it as a community-based innovation contest.

First, I have argued that members of innovation communities belong to different organizational departments or even different organizations (Bansemir et al., 2012; Muhdi & Boutellier, 2011). Students participating at the WIN Innovation Community belong to different study programs. In the case of the WIN Innovation Community, members study in the fields of business administration, economics, international business studies, engineering, economic education, and information systems. As such, the group of students involved in the community is of heterogeneous nature and is comparable to a community that crosses functional boundaries of organizations.

Second, corporate innovation community members work in both offline as well as online settings. The Institute of Information Systems provides an online platform for community members, which allows them to present their ideas as well as to discuss and to evaluate concepts with other members. Yet, the provided platform is not the only interaction channel used by community members. When asked which interaction channels they used in the context of the WIN Innovation Community, students reported that they used a wide range of online channels (e.g., online communities such as facebook or e-mail) as well as offline channels (e.g., team meetings) in varying intensity. Table 30 summarizes the answers of the students with regard to the frequency with which they used different interaction channels. The numbers represent the average answer score (based on a seven point Likert scale). It highlights the online as well as offline interaction among WIN Innovation Community members.

Table 30: Usage of interaction channels in the WIN Innovation Community

Average answer scores of a 7 point Likert scale on frequency of usage of interaction channels	
Social communities	5.85
Offline	5.15
WIN platform	4.06
E-mail	2.93
Phone	2.61
Chats	2.31
Instant Messaging	1.85
Online groups	1.74

Third, the definition presented in part II stresses that community members innovate in a joint effort as opposed to individual accomplishment of tasks (Franke & Shah,

2003). Members of the WIN Innovation Community are assigned to teams of four or five members in order to fulfill the task in the context of the community. In addition, students are encouraged to exchange ideas with community members outside their teams. Hence, innovation activities within the WIN Innovation Community are clearly of interactive and joint nature.

Fourth, my definition of innovation communities requires community members to engage in community activities on a voluntary basis (Fleming & Waguespack, 2007; Franke & Shah, 2003). While some of the activities have to be fulfilled by the students in the course of their studies, a large extent of activities in both offline as well as online channels remains voluntarily. More specifically, students have to upload a solution for the task they received from the Insitute of Information Systems with their team as well as to publish 20 comments on the solutions proposed by other teams on the online platform. Hence, the mandatory activities of students within the community capture only a small fraction of community activities. Moreover, I focus my analysis of social capital on team external ties within the community, which are outside the mandatory activities of community members.

Fifth, my definition of corporate innovation communities focuses on those communities that receive organizational support and distinguishes corporate communities from those that appear self-organized, for example in the context of sports (Franke & Shah, 2003). The WIN Innovation Community receives support from the Institute of Information Systems in numerous ways. For instance, the department provides an online platform where community members can exchange ideas and knowledge. As another example, students have the opportunity to discuss their ideas and concepts with employees from the department in offline settings. Thus, I propose that the WIN Innovation Community meets the five characteristics I presented in part II as a definition of corporate innovation communities. As a result, the WIN Innovation Community represents a fruitful empirical field to test the model presented.

4.2 Data collection approach

To examine the model presented empirically, data from WIN Innovation Community members was collected using an online survey[31]. Online surveys are particularly popular in the field of innovation communities (e.g., Füller, 2006; Jeppesen & Lakhani, 2010; Lakhani & von Hippel, 2003) and have a number of advantages, most importantly with regard to time and money (Lüthje et al., 2005)[32]. All 1,746 members of the WIN Innovation Community were invited by e-mail to participate in the survey. Several steps have been helpful to ensure successful data collection (see Figure 23).

Steps to ensure successful data collection	Related arguments in the literature & consequences
A professor explained the importance of survey participation to the students.	Top management support is associated with high response rates (Moldoveanu & Baum, 2011).
The invitation e-mail to the survey explained the students the goal of the study.	Understanding the research goal increases acceptability of surveys in OSS communities (Balka et al., 2010).
Students had the opportunity to participate in a lottery if they respond to the invitation.	Lotteries among survey participants improve response rates (Hertel et al., 2003).
Participants were ensured confidentiality and anonymity.	Anonymity is associated with higher measurement validity and accuracy of responses (Kratzer et al., 2008).
Different Likert scales, yes/no questions as well as open questions were used.	Different answer types work against common-method bias (de Luca & Atuahene-Gima, 2007).
Short participation time.	Short participation time is beneficial for response rates and survey completion (Balka et al., 2010).

Figure 23: Steps to ensure successful data collection

[31] Other researchers have been involved in the data collection process as well. See annex 1 for further information.

[32] Of course, online questionnaires are also associated with disadvantages. Most significantly, sample biases may be a problem as some members of the population might not have access to the internet (Lüthje et al., 2005). In the case of the WIN Innovation Community, however, each community member had access to the internet – otherwise he or she would not have been able to register on the online platform.

First, a professor explained the importance of survey participation in a lecture before the invitation was sent to the community members. Moldoveanu and Baum (2011) highlighted the importance of top management support in the context of data collection and its implications for high response rates. In the focal empirical setting, I argue that support by a professor is similarly beneficial.

Second, the invitation e-mail itself explained the students the goal of the study presented here. This procedure is also often discussed in extant studies on innovation communities (e.g., Stam, 2009) and social capital (e.g., Stam & Elfring, 2008). For instance, Balka, Raasch, and Herstatt (2010) suggest that it was important to explain the goal of their study to the focal OSS community to increase acceptability of their research efforts among community members.

Third, the e-mail informed the students that they have the opportunity to participate in a lottery if they respond to the invitation. Prizes were Amazon vouchers and USB sticks worth more than 650 Euros in total. A lottery among participants of online surveys is a popular tool among scholars to incentivice participation in the study and, hence, to improve response rates (e.g., Hertel et al., 2003). Because motivations to engage in the online survey are fundamental different from motivations to engage in community activities, I suggest the incentives to participate in the survey do not bias my findings with regard to motivations as antecedents of community social capital.

Fourth, students were assured that data is analyzed with confidentiality and anonymity. Especially in studies including self-ratings, anonymity is associated with higher measurement validity and accuracy of responses (Kratzer, Gemünden, & Lettl, 2008; Un, Cuervo-Cazurra, & Asakawa, 2010). This aspect is of particular importance in the context of this study because participants reported on their own personality, as well as on their social relationships to other community members.

Fifth, the survey was designed in a way that participants had to answer the questions in different ways (i.e., different Likert scales, yes/no questions as well as open questions). This approach is recommended because it works against problems of common method bias (de Luca & Atuahene-Gima, 2007), a threat the study presented here is particularly vulnerable to because both exogeneous and endogeneous variables are measured within one questionnaire (Fey & Birkinshaw, 2005). Jeppesen and Laursen argue that if respondents have "to answer in different ways in various parts of the questionnaire" (2009: 1584), the likelihood of a common method bias is reduced. I tested for common method bias using the Harman's single factor test (Podsakoff & Organ, 1986). An unrotated solution resulted in nine factors and the first factor explained only 25.103% of the variance. Hence, the Harman's single factor test indicates that common method bias is not a major problem in the data collected.

Finally, the e-mail informed the students that participation in the survey will not last longer than 20 minutes. Long participation time is harmful for response rates (and in particular for questionnaire completion) in online surveys and community studies (Balka et al., 2010; Hertel et al., 2003).

Before the survey was sent out, the questionnaire was pretested with two groups of students (n_1=18 and n_2=11) who had participated in the WIN Innovation Community the years before[33]. A number of scholars have accomplished pretests to improve the quality of questionnaires (e.g., Agerfalk & Fitzgerald, 2008; Tortoriello, Reagans, & McEvily, 2012). Similar to Verworn (2009), the three most important goals in pretesting the survey were (1) to verify validity of constructs as well as to adjust the questions, (2) to assess the translation into German, and (3) to improve the quality of the survey as a whole before the actual study starts.

Data was gathered between the 25.01.2012 and the 19.03.2012. In order to increase the response rate further, non-respondents were reminded after 7, 15, 29, and 35 days after the first invitation. Sending out reminders is a popular approach applied by many scholars to increase response rates (e.g., Foss et al., 2010; Ma & Agarwal, 2007; Nambisan & Baron, 2010). At the end, 1,210 community members responded to the invitation. However, 217 participants did not complete the survey and had to be excluded. Thus, 993 complete questionnairs were received, which equal a response rate of 56.87%. A high response rate like the one achieved is the best method to counterveil response biases (Jeppesen & Laursen, 2009). In the words of Armstrong and Overton: "The most commonly recommended protection against nonresponse bias has been the reduction of nonresponse itself" (1977: 396).

[33] Group 1 completed a first version of the survey on paper basis, which included items of personality as well as of network size, trust, and shared visions of community members. The items were adapted based on the findings of the first pretest. Subsequently, Group 2 completed a revised version of the survey online. This second pretest indicated that completion time of the survey was at borderline. In consequence, the survey was shortened where possible.

5 Data analysis approach

This chapter describes the SEM approach applied in this study. I start with a short description of SEM itself and mark out why it is suitable for the research presented in this part. After that, I discuss aspects of exploratory factor analysis (EFA), confirmatory factor analysis (CFA), and model fit, which I apply in the next step to analyze the data gathered at the WIN Innovation Community.

5.1 SEM and its use in this study

Diamantopoulos and Siguaw define structural equations modeling (SEM) as "a multivariate statistical technique which combines (confirmatory) factor analysis and econometric modeling for the purpose of analyzing hypothesized relationships among latent (i.e., unobserved or theoretical) variables measured by manifest (i.e., observed or empirical) indicators" (2000: 4). In a similar vein, Gefen et al. (2011) describe structural equation models as estimates of path diagrams involving latent variables measured with multiple indicators. As such, SEM deals with a number of mathematical models envolving both relationships between latent variables and corresponding indicators as well as relationships between different latent variables, which are estimated simultaneously. In consequence, Hair and co-authors describe SEM as "similar to a series of multiple regression equations" (2010: 634). SEM has its foundations in factor analysis and multiple regression analysis (Hair et al., 2010) and is also known as "path analysis with latent variables"(McDonald & Ho, 2002: 64).

SEMs are typically composed of a (1) measurement as well as a (2) structural model. First, the measurement model describes the relationships between each latent variable (often referred to as constructs) and the corresponding empirical observed variables (often referred to as indicators) (Diamantopoulos & Siguaw, 2000). In contrast, the structural model provides information about how the latent variables are related to each other (Gefen et al., 2011)[34]. The unique characteristic of SEM is the

[34] Sometimes, the literature uses slightly different terms for these sub-models in SEM (e.g. McDonald & Ho, 2002).

simultaneous estimation of parameters of both the measurement as well as the structural model. In other words, SEM assesses both the measurement of latent variables as well as hypothesized causal paths between the latent variables in one step (Gefen et al., 2011).

The literature regularly distinguishes two approaches in structural equations modeling: (1) covariance-based structural equation modeling (CBSEM) and (2) partial least squares (PLS). Each of these approaches is characterized by advantages and disadvantages.

As Gefen et al. (2011) highlight, the strength of PLS lies in modest requirements with regard to sample sizes as well as distributional assumptions, and exploratory research. At the same time, PLS has its drawback because it does not allow scholars to estimate measurement error variances explicitly and "lacks an overall inferential test statistic" (Gefen et al., 2011: v). Hence, PLS does not allow estimation of model fit. In contrast, CBSEM is associated with confirmatory research, and stronger requirements with regard to distributional assumptions and sample sizes. Related to that, Diamantopoulos and Siguaw (2000) highlight the strengths of CBSEM in confirming developed hypotheses of relationships between different theoretical concepts. Diamantopoulos and Siguaw describe CBSEM as an "aggregate methodology" (2000: 5). In other words, CBSEM estimates parameters based on covariance matrices as opposed to individual measures and cases. As the research presented in this part is obviously of confirmatory nature (the model presented is based both on qualitative research findings presented in part III as well as established theory), the use of the CBSEM approach is appropriate.

A three step approach applied to examine the validity of both measurement as well as structural model is summarized in Figure 24. First, an EFA which includes all indicators offers initial insights on the measurement validity of my model. Second, EFA, CFA, and model fit analysis on construct level address the goal to investigate measurement model validity further. Third, I apply model fit indices as well as analysis of path significance to evaluate the structural model. Some methodological issues with regard to EFA, CFA, and model fit indices are presented in the following.

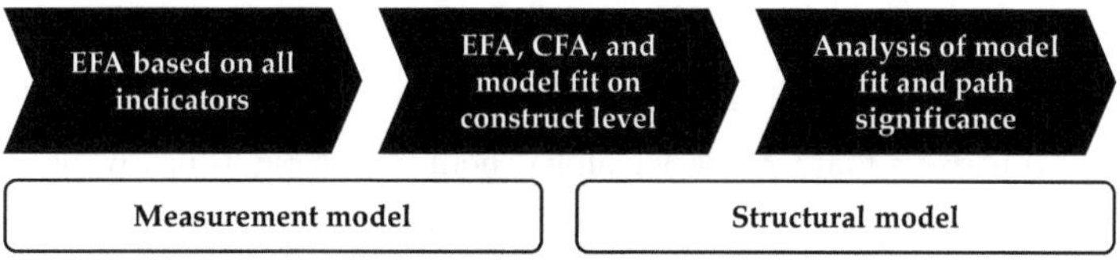

Figure 24: Three step approach to examine validity of measurement and structural model

5.1.1 Exploratory factor analysis

Factor analysis in general is concerned with identifying a number of independent factors out of a higher number of observed indicators (Backhaus, Erichson, Plinke, & Weiber, 2003). EFA can be described by the term "unrestricted" and analyses data without prespecified hypotheses for the structure of the data (Peterson, 2000).

In most cases, the Kaiser criterium is used to determine the number of factors that are extracted out of the data. According to the Kaiser criterium, the number of extracted factors is equal to the number of factors with an eigenvalue higher than one (Tinsley & Tinsley, 1987). Eigenvalues are the sum of the squared factor loadings of one factor over all indicators (Backhaus et al., 2003). In applying EFA, the explained variance of the factors extracted is of major importance. As Tinsley and Tinsley put it: "[A]n analysis in which the factors explain only 30-40% of the estimated common variance obviously leaves an alarming amount of common variance unexplained" (1987: 420). As a result, a common rule of thumb is that extracted factors should explain at least 50% of variance in the data. In addition, factor loadings of the indicators should be higher than 0.4 with regard to the extracted factor that represents the underlying construct (Zinnbauer & Eberl, 2004). The next step, after performing an analysis based on the Kaiser criterium over all indicators, is an analysis of each latent construct that a researcher wants to examine. A first step is to perform several EFAs for each latent construct that the researcher aims to identify. Here, the researcher includes all indicators into the analysis that are supposed to load on one factor with the goal to extract one factor per construct (Weiber & Mühlhaus, 2010).

A second step in this context is reliability analysis based on Cronbach's alpha and item-to-total correlation. Cronbach's alpha, originally introduced by Cronbach

(1951), is a measure of the internal consistency reliability of a factor. The higher the covariances between the indicators are, the higher is the alpha, with a maximum value of one. While there is disagreement on an acceptable value of the alpha, there is some sort of consensus that Cronbach's alpha should be as big as 0.7, while factors with only two or three indicators can also have lower alphas because the value of the alpha increases with the number of indicators (Schnell, Hill, & Esser, 2011). Reliability analysis on the basis of Cronbach's alpha is in many cases complemented with an analysis of item-to-total correlation. Item-to-total correlation represents the correlation of an indicator with the sum of all other indicators. The rule of thumb is that, if Cronbach's alpha is below the desired threshold, items that have an item-to-total correlation below 0.5 are deleted from further analysis (Zinnbauer & Eberl, 2004).

5.1.2 Confirmatory factor analysis

In contrast to EFA, CFA is restricted in a way as it includes hypotheses with regard to the structure of the data (Peterson, 2000). McDonald and Ho highlight the central role of CFA in SEM and argue that "[i]n most applications [of SEM] the measurement model is a conventional confirmatory factor model; the latent variables are just common factors and the error or specific terms are uncorrelated" (2002: 65). The CFA includes, beside the model fit indices discussed in the next section, the analysis of (1) individual item reliability, (2) composite reliability, (3) average variance extracted (AVE), and (4) the Fornell-Larcker-criterium (Bagozzi & Yi, 1988; Zinnbauer & Eberl, 2004).

First, individual item reliability decribes how well an indicator measures a construct (Zinnbauer & Eberl, 2004). Expressed more formally, single item "reliability of a measure is equal to its true score variance divided by the total variance" (Bagozzi & Yi, 1988: 80). An accepted rule of thumb says that 40 to 50% of the variance of an indicator (or item) should be explained by the corresponding latent construct behind the focal indicator (i.e., item or indicator reliability should be 0.4 or 0.5) (Backhaus et al., 2003).

Second, composite reliability is concerned with the reliability of the sum of the indicators for one underlying construct (Weiber & Mühlhaus, 2010). Sometimes researchers explain it as the indicator reliability on construct level and compare composite reliability with Cronbach's alpha. As Ma and Agarwal put it: "Similar to Cronbach's alpha, composite reliability is a measure of internal consistency. Unlike Cronbach's alpha, the composite reliability takes into account the actual loadings used to construct factor scores, and thus is a better measure of internal consistency" (2007: 55). Bagozzi and Yi (1988) argue that composite reliabilities of 0.6 or higher are desirable.

Third, average variance extracted (AVE) describes (similar to composite reliability) how well a construct is measured by all corresponding indicators (e.g., Zinnbauer & Eberl, 2004). While composite or construct reliability indicates the reliability of the unobserved variable, AVE measures "the amount of variance captured by the construct in relation to the amount of variance due to measurement error" (Fornell & Larcker, 1981: 45). AVE is considered to be adequate at a threshold of 0.5 (Bagozzi & Yi, 1988; Fornell & Larcker, 1981).

Fourth, the Fornell-Larcker-criterium is concerned with discriminant validity. Weiber and Mühlhaus (2010) describe discriminant validity as given if measures of different constructs differ significantly. The Fornell-Larcker-criterium (1981) postulates that the ratio between the AVE of a construct and the squared correlation of that construct with every other construct should be higher than one.

5.1.3 Evaluation of model fit

As discussed above, a central characteristic of SEM is the simultaneous estimation of both measurement as well as structural models (Diamantopoulos & Siguaw, 2000). To evaluate the goodness (or badness) of an integrated (measurement and structural) model, SEM researchers use the term of model fit (Weiber & Mühlhaus, 2010). In this context, SEM procedures compare the model-implied covariance matrix with the empirical covariance matrix (Diamantopoulos & Siguaw, 2000) and rely on both statistical tests as well as descriptive model fit indices.

With regard to statistical tests, the χ^2 statistic and the degrees of freedom of the model are of central importance. The χ^2-test tests the null hypothesis that the empirical and the model implied convariance matrixes are equal against the hypothesis that the two matrixes are unequal. If the χ^2-test produces a p-value<0.1, the null hypothesis is rejected and the χ^2-test indicates bad model fit (Weiber & Mühlhaus, 2010). The χ^2 statistic does also allow comparison between alternative models. In this context, the change in χ^2-test proposed by Bentler and Bonett (1980) is applied. The change in χ^2-test looks at the differences between χ^2 statistics and degrees of freedom of the compared models and rejects or accepts the hypothesis that the alternative model is superior to the original model based on the central χ^2 distribution.

However, the χ^2-test is associated with a number of weaknesses (e.g., strong distributional assumptions) and, in consequence, many researchers have questioned its applicability in practice (Zinnbauer & Eberl, 2004). Bearden, Sharma, and Teel (1982) demonstrate in a simulation study that the χ^2-test rejects models too often. A fundamental problem of the χ^2-test is its sensitivity to sample size. In other words, "as the sample size increases, the chances of rejecting a model (whether true or false) also increase. If the sample size is very large, [...] χ^2-test is more likely to reject a false model. But the test is also more likely to reject a true model as well" (Bagozzi & Yi, 1988: 77). Related to that, McDonald and Ho (2002) review extant literature on psychology and SEM and find that out of 41 studies, only five reported a non-significant χ^2. As a result, Schermelleh-Engel, Moosbrugger, and Müller recommend that "not too much emphasis should be placed on the significance of the χ^2 statistic" (2003: 33).

As a reaction to the limitations of the χ^2-test, scholars have developed descriptive indices in order to be able to evaluate model fit in cases where the χ^2-test is not applicable (Weiber & Mühlhaus, 2010). Among others, Gefen and colleagues (2011) stress the strictness of the χ^2-test and propose that researchers should use additional indices to evaluate the fit of their research model. In their words, "to the extent that models are only simplifications of reality, it is likely that sufficient statistical power will tend to lead to a rejection of models according to this strict χ^2-test. Acknowledging this failure of exact fit, researchers are left to determine whether the model may nevertheless represent a useful approximation. This has given rise to a variety of alternative fit indices" (Gefen et al., 2011: ix). In a similar vein, McDonald and Ho argue that "an abundance of indices has been developed as measures of the goodness or badness of the approximation to the distribution from which the sample was drawn" (2002: 72). Scholars have debated for many years about what indices SEM researchers should report. In the words of Schermelleh-Engel et al.: "There is a consensus that one should avoid to report all fit indices that have been developed since the first days of SEM, but there is a certain disagreement on just which fit indices to consider for model evaluation" (2003: 51).

I follow the suggestions of Hu and Bentler (1998, 1999) and distinguish between absolute fit indices and incremental fit indices[35]. Absolute fit indices directly compare the sample covariance matrix with the model-implied covariance matrix (Schermelleh-Engel et al., 2003). I report two indices belonging to this category, namely RMSEA and SRMR. The reporting of RMSEA in combination with SRMR is recommended by Hu and Bentler (1999).

RMSEA stands for Root Mean Square Error of Approximation. The RMSEA is often reported (McDonald & Ho, 2002) and estimates the lack of fit of the model per degree of freedom (Gefen et al., 2011). More formally, the RMSEA is "the square root of the estimated discrepancy due to approximation per degree of freedom" (Schermelleh-Engel et al., 2003: 36). Compared to the χ^2-test, the RMSEA is more sensitive as it assesses approximate model fit in the population (Schermelleh-Engel et al., 2003). Advantages of the RMSEA are relatively independence of sample size and favor of parsimony models (Browne & Cudeck, 1993). While RMSEA was original developed by Steiger and Lind (1980), it is accessible in Browne and Cudeck (1993), who suggest cutoff values of 0.05 for good model fit, 0.08 for acceptable model fit, and argue that values higher than 0.1 are unacceptable. The cutoff values of Browne and Cudeck (1993) are widely accepted among SEM researchers today (e.g., McDonald & Ho, 2002). For instance, Hu and Bentler (1999) recommend rules of thumb of 0.05 for RMSEA when used in combination with SRMR.

SRMR is an abbrevation for Standardized Root Mean Square Residual (Gefen et al., 2011). The unstandardized version (i.e., RMR), "the positive root of the unweighted average of the squared fitted residuals" (Gefen et al., 2011: x), is seldom used because the RMR varies with the scale of the observed variables. SRMR is calculated via a standardized residual matrix, but remains purley descriptive in nature (Schermelleh-Engel et al., 2003). Because of its standardization, the application

35 Other scholars use different terms for the two categories discussed above. For instance, McDonald and Ho (2002) distinguish between absolute and relative fit indices and Schermelleh-Engel et al. (2003) use the terms indices of overall model fit and indices based on model comparison.

of the SRMR is widely accepted with recommended threshold values between 0.05 and 0.10 for good fit (Schermelleh-Engel et al., 2003; Weiber & Mühlhaus, 2010). Hu and Bentler (1999) recommend a cutoff value of 0.06 when used in combination with RMSEA.

In contrast to absolute fit indices, incremental indices (or comparison indices as others have labeled them) compare some form of a baseline model against the model of interest (Schermelleh-Engel et al., 2003). Often, the baseline model is the so called independence model, where all variances are fixed to zero and all factor loadings are fixed to one, leaving the only parameters to be estimated the variances of the variables. Another baseline model is the null model, a model which fixed parameters equal to zero (i.e., no parameters have to be estimated) (McDonald & Ho, 2002). As such, incremental fit indices describe how much a model of interest represents an improvement compared to some form of baseline model. Although reporting of RMSEA and SRMR are sufficient according to the arguments of Hu and Bentler (1999), I complement my reporting with three often used incremental fit indices, i.e., CFI, GFI, and AGFI.

One of the most often reported incremental fit indices is the CFI (Comparative Fit Index) (McDonald & Ho, 2002). Originally introduced by Bentler (1990), CFI uses the independence model as a baseline model, i.e., compares the indepedence model and the model of interest to evaluate model fit (Schermelleh-Engel et al., 2003). One major advantage of the CFI is its robustness against changes in sample sizes (Schermelleh-Engel et al., 2003). CFI ranks from zero to one, with good fit at 0.97 and acceptable fit at 0.95 (Schermelleh-Engel et al., 2003).

The Goodness of Fit Index (GFI) "measures the relative amount of the variances and covariances in the empirical covariance matrix [...] that is predicted by the model-implied covariance matrix" (Schermelleh-Engel et al., 2003: 42). Hence, the GFI baseline model is the null-model with fixed parameters to zero and GFI "measure(s) how much better the model fits as compared to no model at all" (Jöreskog & Sörbom, 1993: 122). Values of GFI rank from zero to one with higher values representing better fit. Rules of thumb for cut off values demand 0.95 for good fit and 0.9 for acceptable fit (McDonald & Ho, 2002; Schermelleh-Engel et al., 2003).

The Adjusted Goodness of Fit Index (AGFI) takes a bias of the GFI associated with model complexity into account (Schermelleh-Engel et al., 2003). To do so, it integrates degress of freedom as well as the number of parameters within the model into its calculation (Jöreskog & Sörbom, 1993; Weiber & Mühlhaus, 2010). In consequence, the AGFI rewards models with less complexity and fewer parameters (Schermelleh-Engel et al., 2003). AGFI is a measure for the explained variance within the model (Zinnbauer & Eberl, 2004) and "indicates the relative amount of variances and covariances jointly accounted for by the hypothesized model" (Bagozzi & Yi, 1988: 79). AGFI is robust against violation of the normality assumption as well as variations in sample sizes and ranks from zero to one, with a rough threshold guideline at 0.9 (Bagozzi & Yi, 1988). Table 31 provides an overview of the indices used.

Table 31: Overview of indices

	Indices	*Threshold*
EFA	Kaiser criterium	≥0.4 factor loadings; ≥50% explained variance
	Cronbach's alpha	≥0.7, but contingent upon number of items
	Item-to-total correlation	≥0.5 (if alpha is below the desired threshold)
CFA	Item reliability	≥0.4
	Composite reliability	≥0.6
	AVE	≥0.5
	Fornell-Larker-ratio	≤1.0
Model fit	χ^2-test	p-value ≥0.1
	RMSEA	≤0.05
	SRMR	≤0.06
	CFI	≥0.95
	GFI	≥0.90
	AGFI	≥0.90

While there is general agreement about the usefulness of the indices presented above, the literature also highlights a number of critics. McDonald and Ho (2002) summarize the criticism in four points: (1) The absence of an empirical or mathematical basis for the use of the fit indices, (2) the nonexistence of a guideline whether to choose between absolute and relative fit indices as well as between different baseline models, (3) the inconsistency between the results out of different indices in combination with a huge number of indices available, and (4) unclear origins of bad model fit as indicated by the indices. Related to that, Gefen et al. emphasize that "it is acceptable that not all fit indices be within these threshold rules of thumb" (2011: x).

6 Measures and measurement validity

In this chapter, I present the measures for both exogeneous (the "Big 5" factors as well as intrinsic and extrinsic motivations) and endogeneous (the three dimensions of social capital) variables. As recommended by a number of scholars (e.g., Foss et al., 2010; Kirsch, Ko, & Haney, 2010), extisting scales were used and adapted to the present context[36]. In addition, items were translated into German[37]. In the following, I present the measures used in this study. Unless otherwise indicated, 7-point Likert scales were used, ranging from strongly disagree to strongly agree.

6.1 Measurement of the FFM

Extant literature provides an impressive range of measures for the FFM. For instance, Goldberg (1992) develops a questionnaire that measures the "Big 5" factors with 100 scales, each representing single trait items. Similarly, Costa and McCrae (1992) present measures of the "Big 5" consisting of a huge number of items (i.e., 240 and 60 items). Saucier (1994) presents a shorter version of the measurement presented by Goldberg (1992) and ends up with 40 "mini-markers", as he calls them.

I have already highlighted the importance of short participation times in online surveys. Hence, such long measures for the "Big 5" were not applicable in the context of the study presented here. Gosling et al. state in a similar direction that online surveys put time constraints on data collecting scholars and argue that in such environments "researchers may be faced with a stark choice of using an extremely brief instrument or using no instrument at all" (2003: 505). To remedy shortcomings of very long measures of the "Big 5", Gosling et al. (2003) present a measure of the "Big 5" which consists of a total of 10 items (i.e., two items per "Big 5" factor). Brown and his co-authors (2002) argue in a similar direction and use a total of 20 items to measure the "Big 5", with varying amounts of items for the five factors. Such short measures of the "Big 5" have recently received increasing attention in the literature

[36] Other researchers have been involved in the measurement process. See annex 1 for further information.

[37] The original wording of the items in German is provided in annex 2.

and have been validated in different languages such as Dutch (Hofmans, Kuppens, & Allik, 2008), German (Muck, Hell, & Gosling, 2007), and Spanish (Romero, Villar, Gómez-Fraguela, & López-Romero, 2012). As extant studies show that shorter scales to measure the "Big 5" produce similar reliability, these shorter scales have been used.

Each factor was measured with three items in order to be able to evaluate the validity of the measures within CFA. In total, the "Big 5" have been measured with 15 items. The items presented by Brown et al. (2002) were used as a starting point and evaluated with respect to translatability into German as well as applicability to WIN Innovation Community members. In addition, Saucier's (1994) and Gosling and other's (2003) measures have been used to develop a comprehensive measurement of the "Big 5" factors. The measures were adapted after the two pretests of the questionnaire. Finally, emotional stability was measured with the three items (1a) "moody more than others", (1b) "temperamental", and (1c) "testy more than others". Introversion was measured with three items, namely (2a) "feel uncomfortable in a group of people", (2b) "prefer to be alone rather than in a large group", and (2c) "shy". For openness to experience, the items (3a) "frequently feel highly creative", (3b) "imaginative", and (3c) "inventive" have been used. Agreeableness was measured with the items (4a) "sympathetic", (4b) "kind to others", and (4c) "caring". Finally, conscientiousness was measured with the items (5a) "precise", (5b) "organized", and (5c) "careful".

6.2 Measurement of extrinsic and intrinsic motivations

In order to measure intrinsic motivation, the scales provided by McLure Wasko and Faraj (2005) and Constant, Sproull, and Kiesler (1996) have been used. Again, extant measures have been adapted to the context of the study presented, ending up with the four items: (1a) "I enjoy being part of the WIN Innovation Community", (1b) "I enjoy developing and improving ideas and concepts with other community members", (1c) "It feels good to help someone in the development of ideas within the community", and (1d) "The most important thing for me is to have fun while working in the community".

While measures for intrinsic motivation from the literature on innovation communities are easily adapted to the present context, measures for extrinsic motivation seem not to fit perfectly to an innovation community comprised of university students. For instance, McLure Wasko and Faraj (2005) measure extrinsic motivation unexceptionally by reputation enhancement within the community. However, extrinsic motivation of students is to a large extent determined by other facets such as grades or passing a course (Deci, 1972). As a result, a measure of extrinsic motivation developed for the context of universities and students has been used. More specifically, items from the Motivated Strategies for Learning Questionnaire (MSLQ)[38] (Pintrich, Smith, Garcia, & Mckeachie, 1993) have been taken. The wording of the items was (Pintrich, Smith, Garcia, and Mckeachie, 1991): (2a) "Getting a good grade in this class is the most satisfying thing for me right now", (2b) "The most important thing for me right now is improving my overall grade point average, so my main concern in this class is getting a good grade", (2c) "If I can, I want to get better grades in this class than most of the other students", and (2d) "I want to do well in this class because it is important to show my ability to my family, friends, employer, or others".

6.3 Measurement of social capital

In this section, I present the measurement procedure with respect to the different dimensions of social capital. I start by describing the measurement of network size (structural dimension), followed by the measurement of shared visions and languages (cognitive dimension) as well as trust and tie strength (relational dimension).

6.3.1 Measurement of the structural dimension of social capital

As numerous authors have done before, the structural dimension of social capital was measured as the number of ties a focal individual has to others (Maurer et al., 2011; McFadyen & Cannella Jr., 2004). This aspect is also referred to as degree centrality

[38] While the MSLQ comprises several other measures, Pintrich, Smith, Garcia, and Mckeachie (1993) highlight that measures of the MSLQ can also be used separately.

(Lechner et al., 2010; Tsai, 2000). Extant literature distinguishes between in-degree and out-degree centrality (Tsai & Ghoshal, 1998). In-degree centrality captures the number of sources a unit of analysis gathers knowledge and ideas from, while the out-degree centrality captures the extent to which others gather knowledge from the focal unit of analysis (Tsai, 2001).

An often applied method to measure degree centrality is called name generator (Lin, 1999). The name generator is an egocentric (as opposed to a sociometric) technique which asks respondents one or more questions to generate names that represent ties within a network or, in the present study, community (Reagans & McEvily, 2003). While some scholars combined two or more name generators (Burt, 2007; Tortoriello & Krackhardt, 2010), the approach of Chua et al. (2008) and Seibert et al. (2001) has been applied in this study. These scholars use a name generator which consists of a single question. Because team members are highly likely to be in contact with one another, I restricted my analysis to team external social capital in order to generate sufficient variance. In addition, this procedure allowed me to focus my research on the voluntary aspects of community work.

More precisely, network size was measured with the following question: "Who are the people outside of your team with whom you have been in contact in the context of the WIN Innovation Community? Name up to 5 contacts"[39]. The measurement approach applied is quite similar to the one proposed by Carroll and Teo (1996), who also restricted their analysis to the first five names mentioned by a respondent[40]. To protect anonymity, students were allowed to limit their answers to first names or initials, a procedure adapted from James (2000). I counted the number of ties a respondent has to individuals outside his or her team to receive a measure of community members' network size. This measurement approach has been applied by other scholars before (e.g., Ahuja, 2000).

[39] Note that the formulation "being in contact with someone" captures both aspects of in-degree and out-degree centrality.

[40] The restriction to five names is a common procedure in the application of the name generator (e.g., Lin, 1999; Reagans & McEvily, 2003).

6.3.2 Measurement of the cognitive dimension of social capital

The cognitive dimension of social capital has been measured with four items. The first two items, (1) "My team-external contacts and I share the same ambitions and vision" and (2) "My team-external contacts and I are enthusiastic about pursuing collective goals and missions", are adapted from Tsai and Ghoshal (1998). However, the conceptual article of Nahapiet and Ghoshal (1998) emphasizes that the cognitive dimension of social capital also captures aspects associated with a shared language. In other words, the cognitive dimension of social capital also captures "shared [...] vocabulary that comes from prior joint experience"(Packalen, 2007: 879). Based on these arguments, two items were added to the measurement presented by Tsai and Ghoshal (1998) in order to capture the whole essence of the cognitive dimension. The items were adapted from Collins and Smith (2006), and had the following wording: (3) "My team-external contacts and I use the same language and vocabulary" and (4) "My team-external contacts and I had no problem when communicating with each other". For those who had no tie in the context of the community outside of their team, I set the missing values to one, indicating the lowest extent to which an individual agrees to the statements presented in the items.

6.3.3 Measurement of the relational dimension of social capital

As highlighted above, my model captures two aspects belonging to the relational dimension of social capital, i.e., trust as well as tie strength (Maurer et al., 2011). I present the measures for these two aspects in the following.

6.3.3.1 Trust

Trust was measured with two items adapted from Tsai and Ghoshal (1998). The online survey presented respondents the list of names they answered to the name generator question presented in section 6.3.1. Respondents answered the following yes/no questions: (1) "Please indicate the team-external contacts you believe you can rely on without any fear that they will take advantage of you?" and (2) "Which of your team-external contacts will always keep the promises they make to you?". Because the amount of names a respondent was confronted with varied contingent upon the names he or she named as relevant contacts in the context of the WIN

Innovation Community, the output data of the online survey entailed missing values. I replaced these missing values with a zero, the same value that represents a non-trusted tie. I counted the names nominated as reliable and promise keeping by each individual as my measure for trust. I standardized the resulting measure by dividing it by five such that the resulting measures range from zero to one.

6.3.3.2 Tie strength

According to Hansen (1999), tie strength captures both closeness as well as frequency of interaction. The measure used was adopted from Lechner and co-authors (2010) and is composed of four items. As was the case for my measure for trust, respondents were presented with the list of names of team-external ties they provided through the name-generator question. They answered the following four questions for each tie mentioned on a 7-point Likert scale: (1) "How frequently have you been in contact with your team-external ties in the context of the WIN Innovation Community?", (2) "How frequently did you interact with your team-external ties for social support?", (3) "How close was the working relationship with your team-external ties?", and (4) "How would you describe the support with your team-external ties?". The Likert scales for question (1) and (2) ranged from "less frequent than once a month" to "every day". For questions (3) and (4), Likert scales ranged from "very distant" to "very close". Again, because the amount of names presented to the respondent was contingent upon the network size of the respondent, I replaced missing values with the value one, representing the weakest answer possible. My measure for an individual's tie strength was the sum of the tie strengths to each of his network contacts divided by 35 (i.e., I standardized it to a scale from zero to one with zero being the lowest and one being the highest possible value).

6.4 Measurement model validity

In this section, I examine the validity of the measurement model[41]. In order to do so, I proceed in two steps. First, I perform an exploratory factor analysis based on the whole dataset. Factors are extracted based on the Kaiser criterium. To examine

[41] The presentation of the measurement model validity is adapted from Danzinger (2010).

whether the data at hand is suitable to perform such an analysis, I conduct the Bartlett test and rely on the Kaiser-Meyer-Olkin-criterium. The Bartlett test examines the hypothesis whether the population the sample stems from is characterized by uncorrelated variables. The Bartlett test assumes normaly distributed variables in the sample as well as an approximately χ^2 distributed test statistic. A test statistic significant at the 5% level is an indicator that data are suitable for EFA.

The Kaiser-Meyer-Olkin criterium calculates a test statistic based on the Anti-Image-correlation matrix, which represents the variance which cannot be explained by the other indicators in a multiple regression analysis (Backhaus et al., 2003). Kaiser, Meyer, and Olkin developed a measure of sampling adequacy (MSA), which describes the extent to which the indicators belong together. As such, it provides estimates whether EFA is meaningful in the context of a dataset at hand. A MSA<0.5 indicates that data is not suitable for EFA, while MSA>0.7, 0.8, and 0.9 are associated with the terms middling, meritorious, and marvelous, respectively.

Within the EFA, I use principal axis factoring as extraction method as I assume that the correlations between the indicators are caused by the extracted factors (Weiber & Mühlhaus, 2010). Furthermore, I use oblimin rotation as the assumption of independence of the constructs is not applicable in the present context (see Backhaus et al. (2003) for a discussion on orthogonal and oblique rotation methods).

Second, I analyze each construct separately. In doing so, I analyze indices associated with EFA (Cronbach's alpha, factor loadings, and item-to-total correlations), CFA (indicator reliability, construct reliability, AVE, and Fornell-Larcker-ratio) and model fit (χ^2, SRMR, RMSEA, CFI, GFI, and AGFI). It is important to note that model fit indices can only be estimated and interpreted usefully for constructs with a minimum of four indicators.

In some cases, I delete indicators based on model fit indices. Gefen et al. (2011) stress that it is important to report the value of fit indices if items are deleted based on them. I follow their suggestion. In this study, I often delete indicators based on bad model fit indicated by RMSEA and AGFI. As already discussed, RMSEA penalizes insufficient parsimony and AGFI punishes model complexity. In the context of this study, deletion of indicators leads to the fact that model fit indices cannot be estimated anymore because only three indicators remain to measure the focal construct.

6.4.1 EFA based on the whole data set

As the Bartlett test (χ^2= 27048.838; df= 561; p=0.000) as well as the Kaiser-Meyer-Olkin criterium (MSA= 0.875) indicate sufficient correlation between the reflective indicators[42], the assumption of multiple indicators seems to be reasonable and data to be suitable for EFA.

Based on the Kaiser criterium using principal axis factoring as well as oblimin-rotation, nine factors have been extracted which explained 68.25% of variance. Factor loadings of the items are presented in Table 32.

[42] In this study, exclusively reflective measurement models are used. For a discussion on the difference between reflective and formative measurement models, see for example Weiber and Mühlhaus (2010).

Table 32: Factor loadings of EFA based on the Kaiser criterium

Items	*Structural & relational SC*	*Cognitive SC*	*Intrinsic motivation*	*Extrinsic motivation*	*Conscientiousness*	*Openess to experience*	*Introversion*	*Emotional stability*	*Agreeableness*
STR_SUM	**.886**	-.071	-.022	-.002	.019	-.008	.030	.019	-.006
TRU_1_STD	**.845**	-.097	-.029	.017	.054	.002	.056	.029	.027
TRU_2_STD	**.819**	-.090	-.032	.022	.027	.002	.038	.010	.039
SW_1_STD	**.964**	.063	.026	-.010	.000	-.017	-.026	-.025	-.007
SW_2_STD	**.902**	.058	.036	-.002	-.048	.008	-.058	-.028	-.042
SW_3_STD	**.921**	.004	.012	-.015	-.029	.016	-.024	-.008	-.011
SW_4_STD	**.957**	.002	-.008	-.004	-.018	.010	-.006	.001	-.007
CGN_1	.005	**-.911**	.018	-.021	-.037	.019	-.006	-.019	-.028
CGN_2	.011	**-.872**	.049	-.032	-.070	.019	-.012	-.028	-.032
CGN_3	.008	**-.933**	.000	.032	.036	-.026	.008	.020	.018
CGN_4	.063	**-.874**	-.027	.037	.048	.001	.006	.027	.019
MOT_INT_1	-.002	.005	**.866**	.020	-.007	.040	.015	.013	-.052
MOT_INT_2	-.022	-.025	**.938**	-.010	-.047	.021	-.013	.012	-.047
MOT_INT_3	.028	-.006	**.841**	-.005	-.010	-.006	.027	.036	.033
MOT_INT_4	.007	-.025	**.666**	.014	.024	.017	.015	-.021	.078
MOT_EX_1	-.001	-.014	-.037	**.849**	.019	.045	.005	.028	.061
MOT_EX_2	-.016	-.085	-.092	**.725**	-.041	.000	-.026	-.017	.069
MOT_EX_3	.005	.036	.011	**.747**	-.076	.038	.021	.039	-.129
MOT_EX_4	.025	.045	.205	**.480**	.033	-.053	-.018	-.066	.018
CON_1	.041	.008	-.049	.072	**-.711**	.073	.082	-.005	.132
CON_2	.028	-.007	.000	.021	**-.907**	-.010	.041	.020	.026
CON_3	-.030	-.030	.059	-.007	**-.735**	-.053	-.086	-.012	.010
OTE_1	-.023	-.024	.039	.017	-.010	**.820**	.023	.004	.011
OTE_2	.021	-.002	.000	-.017	.091	**.853**	.014	-.046	.067
OTE_3	.014	.009	.016	.029	-.058	**.780**	-.020	.038	-.016
EXT_1	-.028	-.007	.006	.022	.033	-.012	**.664**	-.088	.004
EXT_2	-.044	.031	-.014	-.032	-.063	.073	**.510**	-.049	-.178
EXT_3	.024	.001	.038	-.009	.034	-.060	**.757**	.076	.130
EST_1	.029	.028	-.034	-.001	.057	-.026	.160	**-.679**	.012
EST_2	-.001	-.018	.055	-.002	-.008	.118	-.183	**-.643**	.020
EST_3	-.002	-.013	-.063	.013	-.040	-.099	.100	**-.726**	-.047
AGR_1	.012	.033	-.018	-.019	-.105	.059	-.002	.056	**.704**
AGR_2	.024	.047	.005	.014	-.090	.077	-.009	.035	**.732**
AGR_3	-.030	-.062	.048	.024	-.005	-.019	.001	-.061	**.756**

Table 32 shows that most of the items load on the factor that represents the underlying construct with factor loadings ≥ 0.4. The measures for network size (STR_SUM), trust (TRU_1_STD and TRU_2_STD), and tie strength (SW_1_STD, SW_2_STD, SW_3_STD, SW_4_STD) constitute an exception and load on the same factor. As Tsai and Ghoshal (1998) argue that these facets of social capital are highly interrelated, factor loadings do not constitute a major concern. This argument is especially tempting as network size is measured with a single name generator and both tie strength as well as trust are measured based on that name generator. Further investigation, especially issues of discriminant validity as examined by the Fornell-Larcker-ratio, are applied in the next step to take a more nuanced analysis whether the concepts of tie strength and trust empirically represent two distinct constructs. All other items load on the desired factor.

6.4.2 EFA, CFA, and model fit on construct level

In this section, I investigate measurement validity on construct level. For each construct, I discuss results of EFA, CFA, and indices of model fit. Indices that result out of EFA are also referred to as first generation indices, indices that result out of CFA are referred to as second generation indices. In those cases where items have been deleted, I follow the suggestions of Gefen et al. (2011) and report the values of the indices on the basis of which I have deleted items. I start with the exogeneous variables (i.e., aspects of FFM and motivations) and end up with the different facets of social capital (i.e., trust, tie strength as well as shared visions and languages). The validity of the measurement of structural social capital is not examined, as network size was measured with a single indicator[43].

6.4.2.1 Validity of the measurement of the FFM

All factors of the FFM have been measured with three indicators. Hence, model fit indices of SEM cannot be calculated and interpreted usefully for single constructs. As such, I present Cronbach's alpha, composite reliability, average variance extracted (AVE), and the Fornell-Larcker-ratio on construct level as well as indicator reliability,

---

[43] As described in the measurement section, network size was measured by a name generator, which was subsequently transformed into a numeric indicator by counting the names mentioned.

t-values, and item-to-total correlations on indicator level. Table 33 summarizes the results for openness to experience.

Table 33: Measurement validity of the construct "openness to experience"

OTE - Openness to experience				
Description of Indicator	*Mean (standard deviation)*	*Item-to-total correlation*	*Item reliability*	*t-value*
OTE_1: "Frequently feel highly creative"	4.81 (1.419)	0.761	0.718	-
OTE_2: "Imaginative"	4.89 (1.387)	0.763	0.726	28.101***
OTE_3: "Inventive"	5.15 (1.215)	0.717	0.614	26.532***
First generation indices		**Second generation indices (local)**		
Cronbach's alpha	0.865	Comp. reliability	0.87	
		AVE	0.69	
		F-L-ratio	0.17	

As indicated in Table 33, measurement of openness to experience is valid because each of the procedures applied to test measurement validity is within the rule of thumb threshold discussed in the literature.

For agreeableness, the indices and estimations relevant for the evaluation of measurement validity are presented in Table 34. Again, I argue for a valid measurement of the factor agreeableness as all values are within the cutoffs suggested in the literature.

Table 34: Measurement validity of the construct "agreeableness"

AGR - Agreeableness				
Description of Indicator	*Mean (standard deviation)*	*Item-to-total correlation*	*Item reliability*	*t-value*
AGR_1: "Sympathetic"	5.72 (1.127)	0.662	0.587	-
AGR_2: "Kind to others"	6.13 (0.925)	0.694	0.663	21.404***
AGR_3: "Caring"	5.51 (1.229)	0.653	0.559	21.006***
First generation indices		**Second generation indices (local)**		
Cronbach's alpha	0.81	Comp. reliability	0.81	
		AVE	0.59	
		F-L-ratio	0.50	

The estimations to evaluate validity of measurement of emotional stability are presented in Table 35. It shows acceptable values for reliability (Cronbach's alpha and composite reliability above 0.7) as well as discriminant validity (Fornell-Larcker-ratio below one). However, the average variance extracted remains with 0.47 slightly below the desired threshold of 0.5, as does item reliability for EST_2 (0.287).

Nevertheless, I propose that the measurement of emotional stability is not flawed because of three underlying reasons. First, the theoretical basis on which the construct has been developed is strong and many scholars have argued that theoretical arguments should always remain important when judging measurement validity (Bagozzi, 1981; Maurer et al., 2011). Second, and more specifically, I propose that the second item captures an important facet of the construct. As a result, I stand back from deleting this item based on the item-to-total correlation criterium, as other scholars have already done (e.g., Verworn, 2009). Third, many scholars in the literature have argued that values of AVE slightly below the threshold of 0.5 are still acceptable for construct reliability (Carbonell & Rodriguez, 2006; Yli-Renko, Autio, & Sapienza, 2001). Together, I propose that the results indicate valid measurement of emotional stability.

Table 35: Measurement validity of the construct "emotional stability"

EST - Emotional stability				
Description of Indicator	*Mean (standard deviation)*	*Item-to-total correlation*	*Item reliability*	*t-value*
EST_1: "Moody more than others"	3.26 (1.565)	0.568	0.544	-
EST_2: "Temperamental"	4.13 (1.533)	0.453	0.287	13.001***
EST_3: "Testy more than others"	3.08 (1.553)	0.575	0.567	13.318***
First generation indices		**Second generation indices (local)**		
Cronbach's alpha	0.712	Comp. reliability	0.72	
		AVE	0.47	
		F-L-ratio	0.20	

Table 36 shows the results of the analysis of measurement validity for introversion. The results with regard to construct reliability (composite reliability=0.68) and discriminant validity (Fornell-Larcker-ratio=0.2) indicate good validity. While the Cronbach's alpha for introversion is below the often applied 0.7 cutoff, it is still in the acceptable range because many authors have argued that Cronbach's alpha is contingent upon the number of indicators (e.g., Schnell et al., 2011). For instance, Muck et al. (2007) investigate the "Big 5" and end up with all Cronbach's alpha below the common threshold value of 0.7, with agreeableness being the lowest at 0.42. However, in their argument, short measures of the "Big 5" such the ones applied in this study are "designed to optimize content validity; [...] the goal of maximizing content validity inevitably comes at the expense of internal consistency" (Muck et al.,

2007: 168). Even more, Romero et al. (2012) report a Cronbach's Alpha as low as 0.38 for agreeableness.

In addition, the average variance extracted is below the common rule of thumb of 0.5. My arguments for retaining INT_2 (reliability=0.288) are similar to those applied for the previous construct emotional stability. It is important to note that other scholars have also reported values of the AVE that were similar to the results estimated here. For instance, Matzler, Bidmon, and Grabner-Kräuter (2006) report an AVE of 0.43 and Cillo, de Luca, and Troilo (2010) state that an AVE of 0.42 indicates acceptable validity. As a result, I suggest based on the numbers presented in Table 36 that the measurement of introversion can be considered as valid.

Table 36: Measurement validity of the construct "introversion"

INT - Introversion				
Description of Indicator	*Mean (standard deviation)*	*Item-to-total correlation*	*Item reliability*	*t-value*
INT_1: "Feel uncomfortable in a group of people"	2.53 (1.482)	0.529	0.514	-
INT_2: "Prefer to be alone rather than in a large group"	2.63 (1.615)	0.434	0.288	11.628***
INT_3: "Shy"	3.00 (1.610)	0.511	0.466	11.688***
First generation indices		**Second generation indices (local)**		
Cronbach's alpha	0.677	Comp. reliability	0.68	
		AVE	0.41	
		F-L-ratio	0.22	

Table 37 shows the results with regard to the last construct of the Five Factor Model, i.e., conscientiousness. It shows that the numbers indicate acceptable results for reliability (Cronbach's alpha=0.854; composite reliabilty=0.86), average variance extracted (0.67), and discriminant validity (Fornell-Larcker-ratio=0.45).

Table 37: Measurement validity of the construct "conscientiousness"

CON - Conscientiousness				
Description of Indicator	*Mean (standard deviation)*	*Item-to-total correlation*	*Item reliability*	*t-value*
CON_1: "Precise"	5.73 (1.260)	0.714	0.626	-
CON_2: "Organized"	5.54 (1.323)	0.800	0.877	26.840***
CON_3: "Careful"	5.27 (1.476)	0.679	0.535	24.369***
First generation indices		**Second generation indices (local)**		
Cronbach's alpha	0.854	Comp. reliability	0.86	
		AVE	0.67	
		F-L-ratio	0.45	

6.4.2.2 Validity of the measurement of intrinsic and extrinsic motivations

As discussed above, intrinsic and extrinsic motivations were initially measured with four indicators per construct. In examining the measurement validity empirically, inacceptable values of RMSEA and AGFI indicated unsufficent parsimony in the measurement procedure. As a result, I deleted one item per construct and ended up with three indicators for both intrinsic as well as extrinsic motivation.

Table 38 shows indices to evaluate measurement validity for intrinsic motivation. As shown in the Table 38, I deleted MOT_INT_4 because RMSEA (=0.131) indicated insufficient parsimony of the measurement model. Furthermore, MOT_INT_4 had the lowest item-to-total correlation with an absolute value at 0.659. Based on detailed anaylsis of the wording of the items, I propose that the importance of fun within community settings is already captured by the other items, for instance MOT_INT_1. The three-item-measurement model of intrinsic motivation indicates a valid measurement model with Cronbach's alpha (=0.918), composite reliability (=0.92), average variance extracted (=0.79) and Fornell-Larcker-ratio (=0.15) in the recommended ranges. In a nutshell, I propose that the measurement of intrinsic motivation is valid.

Table 38: Measurement validity of the construct "intrinsic motivation"

MOT_INT - Intrinsic motivation				
Description of Indicator	*Mean (standard deviation)*	*Item-to-total correlation*	*Item reliability*	*t-value*
MOT_INT_1: "I enjoy being part of the WIN Innovation Community"	4.75 (1.642)	0.834	0.784	-
MOT_INT_2: "I enjoy developing and improving ideas and concepts with other community members"	4.73 (1.637)	0.886	0.935	43.601***
MOT_INT_3: "It feels good to help someone in the development of ideas within the community"	4.72 (1.622)	0.782	0.662	34.326***
MOT_INT_4: "The most important thing for me is to have fun while working in the community"	5.01 (1.644)	Deleted based on RMSEA and AGFI		
First generation indices		**Second generation indices (local)**		
Cronbach's alpha	0.918	Comp. reliability	0.92	
		AVE	0.79	
		F-L-ratio	0.15	

Table 39 summarizes the results to evaluate measurement model validity of the construct extrinsic motivation. As was the case for intrinsic motivation, I deleted

MOT_EX_4 because model fit indices indicated insufficient parsimony (RMSEA=0.224; AGFI=0.753). I deleted the indicator with the lowest item-to-total correlation, MOT_EX_4, with an absolute value of 0.500. I suggest that the other items already capture the desire of community members to proof individual capabilities to others. The results again indicate excellent reliability (Cronbach's alpha=0.777; composite reliability=0.79), average variance extracted (=0.56) and discriminant validity (Fornell-Larcker-ratio=0.29).

Table 39: Measurement validity of the construct "extrinsic motivation"

MOT_EX - Extrinsic motivation				
Description of Indicator	*Mean (standard deviation)*	*Item-to-total correlation*	*Item reliability*	*t-value*
MOT_EX_1: "Getting a good grade in this class is the most satisfying thing for me right now"	6.47 (0.952)	0.717	0.812	-
MOT_EX_2: "The most important thing for me right now is improving my overall grade point average, so my main concern in this class is getting a good grade"	6.62 (0.806)	0.651	0.596	21.744***
MOT_EX_3: "If I can, I want to get better grades in this class than most of the other students"	5.62 (1.414)	0.602	0.433	19.601***
MOT_EX_4: "I want to do well in this class because it is important to show my ability to my family, friends, employer, or others"	5.26 (1.748)	Deleted based on RMSEA and AGFI		
First generation indices		**Second generation indices (local)**		
Cronbach's alpha	0.777	Comp.reliability	0.79	
		AVE	0.56	
		F-L-ratio	0.29	

6.4.2.3 Validity of the measurement of social capital

In this section, I present the results of the evaluation of measurement validity of social capital. Validity of the measurement of network size (structural social capital) is not discussed as it has been measured with a single indicator based on the arguments of Seibert et al. (2001). I start with the measurement of the relational dimension of social capital (i.e., trust as well as tie strength), which has been measured based on the name generator and end up with the measurement of the cognitive dimension of social capital. In order to analyse the measurement validity of those constructs that have been measured based on the name generator, I transposed the data in a way that each row represented the answers of a respondent with regard to one network tie. As a result, the answers of each respondent were five times higher than the n. Table 40 summarizes my results with regard to the measurement of trust.

Table 40: Measurement validity of the construct "trust"

SC_TRU - Social Capital: Trust				
Description of Indicator	*Mean (standard deviation)*	*Item-to-total correlation*	*Item reliability*	*t-value*
SC_TRU_1: "Please indicate the team-external contacts you believe you can rely on without any fear that they will take advantage of you?"	0.28 (0.448)	0.911	-	-
SC_TRU_2: "Which of your team-external contacts will always keep the promises they make to you?"	0.26 (0.439)	0.911	-	-
First generation indices		**Second generation indices (local)**		
Cronbach's alpha	0.953	Comp. reliability	0.96	
		AVE	0.93	
		F-L-ratio	0.82	

While all items reported in Table 40 indicate acceptable values (Cronbach's alpha=0.953, composite reliality=0.96, AVE=0.93, and Fornell-Larcker-ratio=0.82), conclusive evaluation of measurement validity based on these numbers is not feasible. For instance, while many researchers report Cronbach's alpha for two item constructs (e.g., Füller, 2006; Muck et al., 2007), Agerfalk and Fitzgerald (2008) omit the reporting of the alpha for constructs measured by two indicators. Because the literature also provides measures with more items, trust was measured with additional items proposed by de Jong and Elfring (2010). This procedure allowed me to benefit from the strengths of the name generator and to evaluate the validity of the measurement of trust more conclusively. The detailed analysis is provided in annex 3 and shows that the correlation coefficient of the trust construct measured based on the name generator and the construct of trust measured similar to de Jong and Elfring (2010) is 0.74, which indicates that both constructs represent similar phenomena.

The results for the measurement validity of tie strength are summarized in Table 41. The initial measurement model, which was composed of four indicators, was characterized by a lack of parsimony (RMSEA=0.241) and exaggerated complexity (AGFI=0.708). Hence, I deleted the item with the lowest item-to-total correlation, SC_SW_2. The second item seems to be very similar to the first item, i.e., being in contact and supporting each other seem to be just about equal meanings in innovation community settings. Based on the numbers presented, I propose that my analysis indicated high measurement validity of the construct tie strength.

Table 41: Measurement validity of the construct "tie strength"

SC_SW: Social capital: Tie strength				
Description of Indicator	*Mean (standard deviation)*	*Item-to-total correlation*	*Item reliability*	*t-value*
SC_SW_1: "How frequently have you been in contact with your team-external ties in the context of the WIN Innovation Community?"	2.21 (1.990)	0.887	0.816	-
SC_SW_2: "How frequently did you interact with your team-external ties for social support?"	1.99 (1.792)	Deleted based on RMSEA and AGFI		
SC_SW_3: "How close was the working relationship with your team-external ties?"	2.16 (1.959)	0.927	0.918	121.920***
SC_SW_4: "How would you describe the support with your team-external ties?"	2.27 (2.049)	0.934	0.941	125.762***
First generation indices		**Second generation indices (local)**		
Cronbach's alpha	0.96	Comp. reliability	0.96	
		AVE	0.89	
		F-L-ratio	0.86	

Finally, Table 42 summarizes my analysis of measurement validity of the cognitive dimension of social capital. Similarly to the measurement of tie strength, the initial four indicator measurement model was characterized by a lack of parsimony (RMSEA=0.597) and inordinate complexity (AGFI=-0.117). Hence, I deleted SC_CGN_2 that had the lowest item-to-total correlation with 0.859. Referring to the wording of the items, I suggest that in hierarchy free environments, ambitions and visions are very similar to goals and missions as there is no formal institution, which formulates a desired shape of community activities. Analysis of the three indicator model indicates valid measurement validity on the basis of Cronbach's alpha (=0.944), composite reliability (=0.95), average variance extracted (=0.88), and Fornell-Larcker-ratio (=0.56).

Table 42: Measurement validity of the construct "shared visions and languages"

SC_CGN: Social capital: Shared visions and languages				
Description of Indicator	*Mean (standard deviation)*	*Item-to-total correlation*	*Item reliability*	*t-value*
SC_CGN_1: "My team-external contacts and I share the same ambitions and vision"	3.57 (2.131)	0.817	0.689	-
SC_CGN_2: "My team-external contacts and I are enthusiastic about pursuing collective goals and missions"	3.44 (2.090)	Deleted based on RMSEA and AGFI		
SC_CGN_3: "My team-external contacts and I use the same language and vocabulary"	4.31 (2.445)	0.933	0.964	43.228***
SC_CGN_4: "My team-external contacts and I had no problem when communicating with each other"	4.63 (2.584)	0.917	0.920	42.167***
First generation indices		**Second generation indices (local)**		
Cronbach's alpha	0.944	Comp. reliability	0.95	
		AVE	0.88	
		F-L-ratio	0.56	

7 Hypotheses testing

In this chapter, I analyse both fit of the overall model as well as significance of paths within the model, which represents an empirical investigation for support of the hypotheses presented earlier. While software packages available for CBSEM (e.g., AMOS) provide a multitude of estimation methods to researchers, I use maximum likelihood because of three underlying reasons. First, maximum likelihood has advantages of a formal statistical test of model fit (Schermelleh-Engel et al., 2003). Second, maximum likelihood seems to be relatively robust against violations of the most often discussed underlying assumption (i.e., multivariate normality) (McDonald & Ho, 2002; Schermelleh-Engel et al., 2003). Third, maximum likelihood is the most often used estimation method in SEM. For instance, Gefen et al. (2011) discuss the advantages of maximum likelihood and argue that it does not raise too much questions from readers because of its familiarity.

To evaluate fit of the overall model, I apply the model fit indices discussed above[44]. The χ^2 of my model is 1080.857 with 370 degrees of freedom (df). The χ^2-test is highly significant (p=0.000). That is, based on the criterium of the χ^2-test, my model would be rejected. Yet, I have discussed the limitations of the χ^2 statistic above, and apply additional indices to evaluate model fit further. Both absolute (SRMR=0.0422; RMSEA=0.044) as well as incremental fit indeces (CFI=0.968; GFI=0.931; AGFI=0.913) indicate a very good fit. Hence, I conclude that my model generally fits the data well.

Figure 25 depicts the model with the maximum likelihood parameter estimates. Standardized parameter estimates show that 10 of my 15 proposed links were significant and in the hypothized direction. Hypothesis 1 relates structural social capital to a) cognitive social capital as well as the two facets of relational social capital, i.e., b) trust and c) tie strength. The results indicate support for H1a (b=0.69; p<0.001), H1b (b=0.52; p<0.001), as well as H1c (b=0.83; p<0.001). Fully supporting Hypotheses 2a (b=0.12; p<0.001) and 2b (b=0.10; p<0.001), respondents who reported greater cognitive social capital also reported more trusted as well as stronger

[44] The empirical covariance matrix is provided in annex 4.

relationships. Finally, Hypothesis 3 relates tie strength to trust and is supported ($b= 0.33; p<0.001$).

Hypotheses 4 and 5 relate openness to experience and agreeableness to the structural dimension of social capital. I find no support for (H4) the relationship between openness to experience and structural social capital ($b=-0.02; p=0.600$) and (H5) the relationship between agreeableness and structural social capital ($b=0.08; p=0.082$). I found marginal support for Hypothesis 6, as the path between emotional stability and structural social capital was positive and significant ($b=0.08; p=0.053$). Hypothesis 7 negatively relates introversion to the structural dimension of social capital. I find empirical support for this hypothesis, as is reflected in a negative, significant path coefficient ($b=-0.12; p=0.003$). In Hypotheses 8, I relate conscientiousness to both (a) structural as well as (b) cognitive social capital. While I find no support for H8a ($b=-0.01; p=0.885$), my data supports H8b ($b=0.06; p=0.019$).

My results fail to support Hypothesis 9 ($b=0.06; p=0.122$), which relates extrinsic motivation to structural social capital. Hypotheses 10a and 10b relate intrinsic motivation to structural as well as cognitive social capital, respectively. My data fails to support H10a ($b=0.06; p=0.106$), but supports H10b ($b=0.05; p=0.049$). The explained variance of the construct of the structural (network size) as well as the cognitive (shared visions and languages) dimension of social capital is 4% and 49%, respectively. The explained variance of the trust construct is 84%, and of the tie strength construct 81%.

I used the change of χ^2-test (Bentler & Bonett, 1980) to compare my hypothesized model to two partly mediated models as well as to three non-mediated models. A similar procedure has been applied by Seibert et al. (2001). The first comparison was between the hypothesized model and a partially mediated model #1, which included paths of the hypothesized model as well as paths between the "Big 5" and the intrinsic and extrinsic motivation and the tie strength construct. The change in χ^2-test indicated that partially mediated model #1 was significantly better than my hypothesized model ($\Delta\chi^2=21.349$; $\Delta df=7$; $p<0.01$). The significance levels of all relationships of the hypothesized model remained equal[45]. Because partially mediated

[45] Although not hypothesized, this model indicated an additional significant relationship between introversion and tie strength ($b=-0.06; p=0.004$).

model #1 was better fitting than my hypothesized model, I compared this model to a second partially mediated model #2. Partially mediated model #2 included, in addition to the relationships investigated at partially mediated model #1, relationships between the exogeneous constructs (i.e., "Big 5" as well as intrinsic and extrinsic motivations) and the trust construct. This model was not significantly better than the partially mediated model #1. I also compared partially mediated model #2 to three non-mediated models. Non-mediated model #1 captured all relationships included in partially mediated model #2 except the relationship between tie strength and trust. Non-mediated model #2 captured all relationships included in non-mediated model #1 except the relationship between shared visions and tie strength as well as trust. Non-mediated model #3 was equal to non-mediated model #2 except missing linkages between network size and shared visions, tie strength as well as trust. None of these non-mediated models fit the data better than partially mediated model #2. Table 43 summarizes my arguments.

Table 43: Change in χ^2-tests of alternative models

Model	$\chi^2(df)$	$\Delta\chi^2(\Delta df)$	*Model comparison*	*Better fit*
Hypothesized	1080.857 (370)			
Partially mediated #1	1059.508 (363)	21.349** (7)	Partially mediated #1 compared to hypothesized	Partially mediated #1
Partially mediated #2	1055.015 (356)	4.493 (7)	Partially mediated #2 compared to Partially mediated #1	Partially mediated #1
Non-mediated #1	1137.999 (357)	-82.98** (-1)	Partially mediated #2 compared to non-mediated #1	Partially mediated #2
Non-mediated #2	1206.769 (359)	-151.754** (-3)	Partially mediated #2 compared to non mediated #2	Partially mediated #2
Non-mediated #3	2097.805 (362)	-1042.790** (-6)	Partially mediated #2 compared to non-mediated #3	Partially mediated #2

**p < 0.01

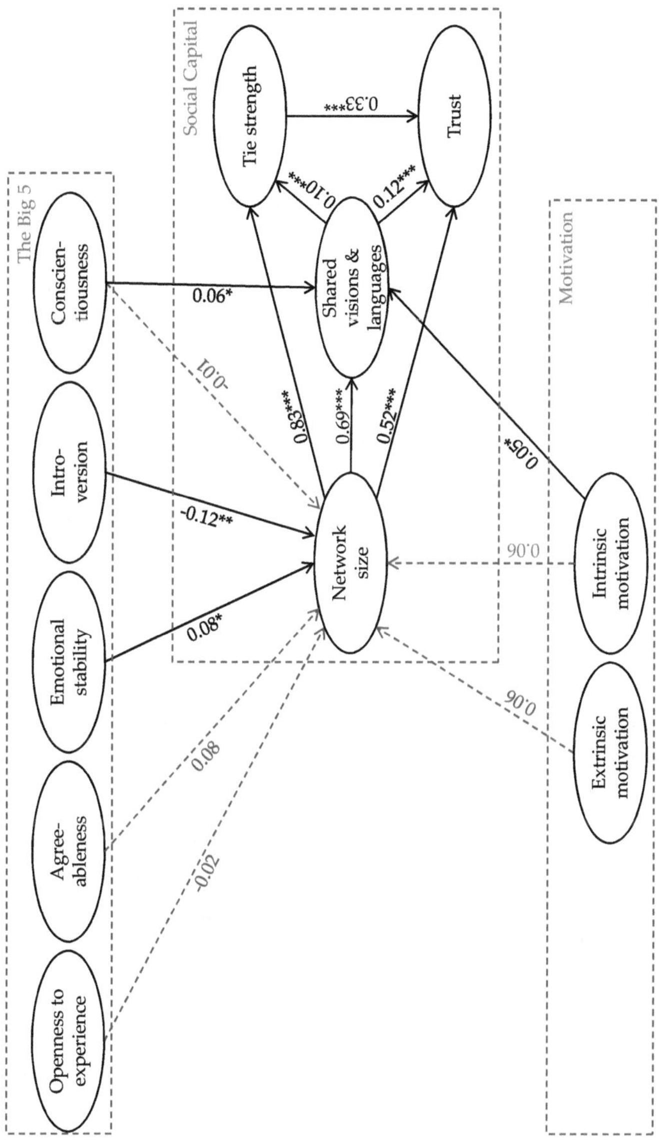

Fit: χ^2=1080.857; df= 370; p<0.001; SRMR=0.0422;RMSEA=0.044; CFI=0.968; GFI=0.931; AGFI=0.913. Parameter estimates are from the completely standardized solution. Significance is indicated by * (5% level), ** (1% level), and *** (0.1% level). Significant relationships are furthermore indicated by bold arrows, insignificant relationships are indicated by light arrows. The measurement models are not shown for the sake of clarity.

Figure 25: Hypothesized model and maximum likelihood parameter estimates

8 Discussion

The study presented in this part was motivated by important limitations in the academic literature on social capital and innovation communities. The model presented builds upon findings derived out of multiple case studies presented in part III as well as theoretical arguments of extant literature on social capital and innovation communities. More specifically, while social capital theorists have advocated for the importance of personality as an antecedent of social capital (Klein et al., 2004), there is, to the best of my knowledge, no study that tests the relationship between personality and the interrelated dimensions of social capital in innovation communities on a quantitative basis. Furthermore, as community members operate on a voluntary basis, there is consensus that both intrinsic as well as extrinsic motivations are important facets in innovation communities (Janzik, 2010). As qualitative analysis in part III revealed, the extent to which members build social networks within innovation communities is contingent upon these motivational aspects. Again, my review of the literature did not identify a study that already tested this relationship quantitatively. The research presented in this part follows the call of Payne et al. (2011) who stress the need for more research on antecedents of social capital. In addition, the model presented takes interrelations between different dimensions of social capital into account, a facet often neglected in the past (Inkpen & Tsang, 2005).

I have integrated literatures of personality psychology, social capital, and innovation communities to propose a model that conceptualizes personality and motivational aspects as antecedents of different dimensions of social capital. My findings are at the same time promising and cautionary: I find empirical support for 10 of my 15 hypotheses. Further, while my model is able to explain high amounts of variance with regard to three facets of social capital (shared visions and languages 49%; trust 82%, and tie strength 83%), the amount of variance of network size explained by the model remains fairly low (4%).

My results demonstrate that two aspects of personality (i.e., introversion and emotional stability) relate to the focal community member's network size. More specifically, introversion affects network size negatively, while emotional stability is positively related to network size. In addition, conscientiousness as well as intrinsic motivation affect shared visions and languages in community settings. Network size of community members was in turn positively related to shared visions, trust, as well as tie strength. Shared visions related positively to trust and tie strength, and tie strength was positively associated with trust.

Empirical data did not support several of my hypotheses with regard to antecedents of network size of innovation community members. For instance, I found no support for a relationship between agreeableness and network size. This evidence is in line with Wolff and Kim (2012) who find that agreeableness does not affect networking across boundaries and that agreeable people engage heavily in maintaining social relationships, but not in creating new relationships. Interpretation of data suggests that the pleasant nature of agreeable community members (Brown et al., 2002) seems not to be sufficient to result in interaction with many peers and in a high number of ties within the community. At the same time, I found no support for openness to experience as an antecedent of network size. I propose that this finding is related to the multifaceted nature of the factor itself (Judge et al., 2002). Similarly, Wanberg et al. (2000) found only marginal support for their hypotheses that openness to experience is related to intense networking behavior. The task-oriented nature of conscientious personalities might explain the insignificant relationship between conscientiousness and network size of community members. As Wolff and Kim (2012) have argued, a narrow focus on the task itself may hinder networking activities. In essence, efficiency orientation is often supposed to be harmful to innovation, especially in the early phases of the innovation process. I furthermore found no significant relationship between extrinsic motivation and community network size. The missing link between these two constructs might be explained by the fear of extrinsically motivated community members to earn disadvantages when collaborating with other community members (Frey & Lüthje, 2011). Two lines of arguments might explain the insignificant relationship between intrinsic motivation and network size. First, high intrinsic motivation can be associated with reduced willingness to compromise (Faraj et al., 2011) and, hence, less interaction with others. The willingness to discuss and accept ideas of others is highly important in innovation community settings. Furthermore, not all facets of intrinsic motivation entail aspects of exchange with other community members. For instance, pure interest in the task of the community (e.g., Lerner & Tirole, 2002) does not necessarily lead to more interaction with other community members. Again, I discuss theoretical implications as well as limitations that are specific to the study presented in this part in the following. I discuss managerial implications of the research presented in this dissertation in part V.

8.1 Theoretical implications

I add to both literatures of social capital as well as innovation communities. I start by discussing my contribution to social capital literature, which is twofold. First, the results presented here provide support for the importance of personality and intrinsic motivation as antecedents of social capital in corporate innovation community settings. Previous research has already demonstrated the importance of personality aspects as predictors for single dimensions of social capital (e.g., Klein et al., 2004). The current study differs from these earlier research attempts and is unique for its inclusion of various dimensions in the analysis of antecedents of social capital in innovation communities. Specifically, the model presented in this part of the dissertation shows how network size as well as shared visions and languages mediate the relationship between personality as well as intrinsic motivation and relational social capital (i.e., tie strength and trust) between community members.

Second, the model presented demonstrates how different dimensions of social capital are related to each other and which dimensions represent levers to influence social capital. While Tsai and Ghoshal (1998) have already proposed a model that looks at the relationships between structural, cognitive, and relational social capital, I take a more nuanced view of the relational dimension. Defined as relationship quality (Yli-Renko et al., 2001), many scholars have treated trust as a synonym of the relational dimension. Another aspect that is also part of the quality of relationships and often discussed in the literature is tie strength, i.e., the closeness and frequency of interaction (Hansen, 1999; Lechner et al., 2010). Similar to Levin and Cross (2004), I treat tie strength as an antecedent of trust.

By proposing a model of antecedents of corporate innovation community social capital, I also contribute to the literature on innovation communities. Because of the voluntary nature, many scholars have studied motivations of community members to participate in community activities (Lerner & Tirole, 2002; Muhdi & Boutellier, 2011). At the same time, scholars have emphasized the importance of exchange and social relationships in such settings (e.g., Dahlander & Frederiksen, 2012). Integrating aspects of motivation as well as social relationships and their quality allows a more detailed investigation of mechanisms at work. In more detail, a social capital lens adds to studies on motivations of community members by investigating the consequences of extrinsic and intrinsic motivations with regard to exchange behavior. Empirical analysis did not confirm extrinsic motivation as an antecedent of social capital. In other words, extrinsic motivation does not affect voluntary exchange with other community members (structural dimension of social capital). If exchange between community members is a central goal and characteristic of innovation communities, this is an alarming finding. To the contrary, intrinsic motivation is positively related to cognitive social capital, which in turn affects tie strength as well as trust. In consequence, intrinsic motivation is beneficial for close, frequent, and trusted exchange within the community.

8.2 Limitations and avenues for future research

Although its large sample represents a strength of this study, the empirical field in which the data was collected at is related to several limitations. For instance, as respondents of the survey were students, transferability to business contexts remains up to future research efforts. Yet, I argue that because of the research question addressed, my findings are sufficently general. More specifically, the "Big 5" personality traits are relatively stable over long periods of lifetime (Judge et al., 1999). At the same time, there are no tremendous differences with regard to the construct of intrinsic motivation in community work between students and other people. To the contrary, extrinsic motivation of students is characterized mainly by grades (e.g., Deci, 1972), while community members in business contexts are motivated differently (e.g., aspects of reputation or usage of the product or service might be more important). Hence, I call for future research that tests the relationship of the presented model in business contexts.

A second limitation of the model presented in this part is that it explains a relatively little amount of variance of the structural dimension of social capital (4%).

This finding is related to the work of Klein and colleagues (2004) who examine personality as an antecedent of centrality in advice and friendship networks. These scholars find that personality explains only 2% of the variance in both types of networks and conclude that "although the individual characteristics [...] do appear to play a role in determining individuals' locations in their [...] networks, they clearly do not tell the entire story" (Klein et al., 2004: 960). I propose that future research is necessary to explore further antecedents of structural social capital in corporate innovation communities. In the context of this dissertation, I propose that a promising step for further studies would be to look at antecedents on the other two levels of the framework proposed in part III. Yet, as the explained variance of the other endogeneous variables of the model presented is sufficiently high, I propose that the contribution of this study remains significant.

V

Discussion and conclusion:

Reflections on research presented in this dissertation

1 Summary of studies

In this dissertation, I have investigated antecedents of social capital in corporate innovation community settings. While extant studies often highlight the importance of communities in innovation processes (Judge et al., 1997; Sawhney & Prandelli, 2000) as well as the beneficial consequences of social capital in such settings (Dahlander & Frederiksen, 2012; McLure Wasko & Faraj, 2005), the literature to date lacks with respect to studies on the antecedents of social capital (Payne et al., 2011). To remedy this research gap, I have incorporated the complex nature of social capital and have considered both multiple dimensions (Nahapiet & Ghoshal, 1998) as well as multiple levels of social capital (Payne et al., 2011). Through this theoretical lens, I have conducted two empirical studies within corporate innovation communities, one multiple case study (Part III) and one SEM approach (Part IV).

1.1 Summary of the study presented in part III

The multiple case study (Eisenhardt, 1989) presented in part III of this dissertation is explorative and triangulates multiple data sources to present a framework of antecedents of corporate innovation community social capital. The framework results out of analysis of data gathered in five corporate innovation communities and offers two important insights. First, innovation community social capital reinforces itself as different dimensions of the concept are highly interrelated. Data indicated that trust and shared visions develop as interaction through network ties between community members takes place, i.e., structural social capital affects cognitive as well as relational social capital positively. Furthermore, shared visions within the community lead to close and trusted interaction. In turn, empirical observations suggested that trust as well as shared visions were associated with further interaction ties within the community, leading to a self-reinforcing cycle of interrelations between different dimensions of social capital in corporate innovation community settings.

Second, the framework presented in part III categorizes antecedents of social capital on three interrelated levels (i.e., individual, organizational, and community level). In the framework offered, the *individual level* captures antecedents of social capital that are specific to individual community members. In this category, empirical evidence suggested that personality (investigated using the FFM) as well as extrinsic and intrinsic motivations of community members affect structural as well as cognitive social capital. In more detail, some community members have been more successful in building social capital than others because they possessed personality traits that allowed them to develop many ties (e.g., the trait of extraversion) and a shared vision (e.g., the trait of conscientiousness) within the community. In addition, the motivations of community members to engage voluntarily in community work affected the extent to which community members developed social capital. While both intrinsic as well as extrinsic aspects were associated with a higher number of community ties, intrinsic motivation also affected the development of shared visions. In essence, because intrinsically motivated community members identify heavily with the community task, they perceive the desired outcome of that task as a vision, which they share with other community members.

Organizational level antecedents comprise antecedents that affect all members of the organization that supports community activities. These antecedents determine the environment in which the community operates to a large extent. The framework presented suggests that an organizational climate that stresses collaboration as well as innovative activities is beneficial for structural (i.e., number of network ties) as well as cognitive community social capital (i.e., shared visions and languages). 3M has successfully established such climates, for instance through allowing a large number of employees to spend a certain amount of working hours on innovative activities as well as knowledge and idea exchange (Studt, 2003). In addition, empirical data indicated that formal (i.e., hierarchy, processes, or projects) and informal organizations affect structural social capital (interaction ties between community members) of corporate innovation communities. Cisco experienced a process that is related to my findings. At Cisco, Gulati and Puranam (2009) studied a reorganization process and observed that the old formal organization remained within the organization as an informal network. Similarly, corporate innovation community members convey existing network structures to community settings and make use of them during innovation processes.

Finally, *community level* antecedents affect the corporate innovation community as a collective. On this level, empirical evidence pointed to three aspects. First, data suggested an inverted U-shaped relationship between creative task design and cognitive community social capital. In more detail, because creative tasks have both benefits (e.g., community members identify with such activities) as well as drawbacks (e.g., community members struggle to agree on the desired outcome of their work), a mediocre level of creativity is most constructive to cognitive social capital. Second, data analysis pointed to the beneficial role of appropriate leadership for cognitive social capital. Leaders who structure community work via workshop preparation or summary of preliminary results foster shared visions within the community because structuring of extant ideas allows community members to make sense of a future outcome of community work. Third, the design of community workshops and platforms affects structural and cognitive social capital in corporate innovation communities. In essence, both offline and online environments support interaction ties in the community as they provide opportunities to exchange knowledge and ideas with other members. In addition, the methods applied during workshops as well as the user interface of community platforms are able to raise shared visions among community members because such aspects are associated with a shared identity.

1.2 Summary of the study presented in part IV

The CBSEM (Diamantopoulos & Siguaw, 2000; Gefen et al., 2011) presented in part IV is of confirmative nature and aims to confirm findings of part III with respect to the reinforcing nature of social capital as well as individual level antecedents. The empirical setting of this study is the WIN Innovation Community, an innovation community of bachelor students at FAU (n=993). I find support for my hypotheses that network size (structural social capital) antecedes shared visions and languages (cognitive social capital) as well as tie strength and trust (relational social capital). In addition, the relationship between network size and tie strength as well as trust is partially mediated by shared visions and languages within the community and tie strength is positively related to trust. Furthermore, I find empirical evidence that personality traits of introversion and emotional stability affect network size – the former negatively and the latter positively. In addition, the personality trait of conscientiousness as well as intrinsic motivation affect shared visions and languages with other community members positively. Figure 26 summarizes empirical findings of both studies presented in this dissertation.

Summary of findings presented in part III

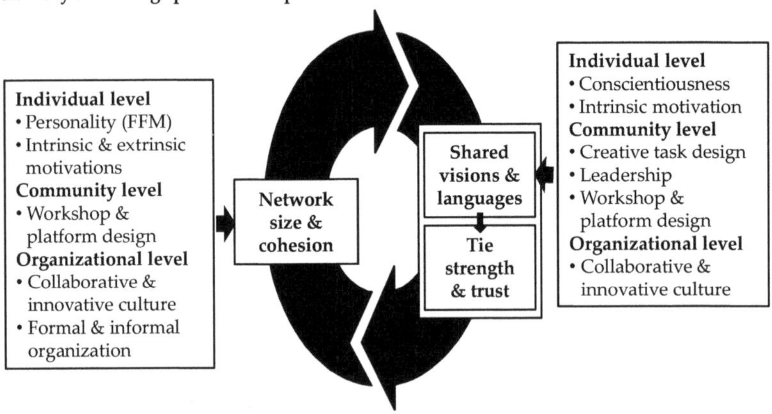

Summary of findings presented in part IV

#	Relationship	Finding
H1a	Network size → Shared visions & languages	Supported
H1b	Network size → Trust	Supported
H1c	Network size → Tie strength	Supported
H2a	Shared visions & languages → Trust	Supported
H2b	Shared visions & languages → Tie strength	Supported
H3	Tie strength → Trust	Supported
H4	Openness to experience → Network size	Not supported
H5	Agreeableness → Network size	Not supported
H6	Emotional stability → Network size	Supported
H7	Introversion → Network size	Supported
H8a	Conscientiousness → Network size	Not supported
H8b	Conscientiousness → Shared visions & languages	Supported
H9	Extrinsic motivation → Network size	Not supported
H10a	Intrinsic motivation → Network size	Not supported
H10b	Intrinsic motivation → Shared visions & languages	Supported

Figure 26: Summary of research findings

2 Theoretical implications

Altogether, the theoretical contribution of this dissertation to the literature on innovation communities and social capital is fourfold. This research contributes (1) to the understanding of the self-reinforcing nature of social capital, (2) to scientific knowledge with regard to the role of individual aspects as antecedents of social capital in community settings, (3) to the role of the organizational environment of corporate innovation communities in the creation of social capital, and (4) to the meaning of the configuration of community activities as antecedents of network structures as well as shared visions and languages.

First, this research provides empirical evidence for a *self-reinforcing cycle of different dimensions* of social capital in corporate innovation communities. To date, the literature lacks with regard to studies that investigate all three dimensions simultaneously – especially the cognitive dimension of the concept has suffered from insufficient consideration (Bolino et al., 2002; Hughes & Perrons, 2011). In addition, extant studies have often neglected interrelations between the different dimensions of social capital (Inkpen & Tsang, 2005). The research presented in this dissertation contributes to the closure of this research gap. In more detail, empirical evidence suggests that interrelations between structural, relational, and cognitive community social capital can be described as a self-reinforcing cycle where interaction between community members provides opportunities to build close and trusted ties as well as shared visions. In turn, close and trusted ties and shared visions are positively related to further interaction within the community. My findings in this context add to those of other social capital scholars who have argued for a reinforcing nature of social capital over time (e.g., Dokko & Rosenkopf, 2010). Yet, to the best of my knowledge, the study presented here is the first to show how this self-reinforcing nature of social capital can be explained by a circle of interrelations between different dimensions of corporate innovation community social capital.

Second, the studies presented offer empirical evidence for the *relevance of individual level antecedents* of social capital in corporate innovation communities. As already mentioned, there is empirical evidence for the relevance of (a) personality and (b) motivations of community members as antecedents of innovation community social capital. A number of scholars have already shown that personality traits significantly influence related concepts (e.g., networking behavior) (e.g., Wolff & Kim, 2012) or single dimensions of social capital (e.g., Klein et al., 2004). I add to the arguments of these researchers as I consider relationships between personality and both structural as well as cognitive social capital. In addition, the research presented explains how structural as well as cognitive social capital mediate the relationship between personality traits and relational social capital within corporate innovation communities.

In addition to personality, the studies presented underpin the relevance of extrinsic and intrinsic motivations as antecedents of community social capital. Related to the voluntary nature of community participation, numerous studies have elaborated on the motivations of members to participate in community activities (Lerner & Tirole, 2002; Muhdi & Boutellier, 2011). Often, scholars have distinguished between extrinsic and intrinsic motivations in this context (Janzik, 2010). I add to the arguments of extant scholars as I investigate the relevance of both extrinsic and intrinsic motivations through the theoretical lens of social capital. While extant studies often restrict their analysis to the decision of community members to participate or not, my approach allows me to take a more nuanced picture of interaction patterns of extrinsically and intrinsically motivated community members. While qualitative data indicated the relevance of both extrinsic and intrinsic motivations as antecedents of social capital in corporate innovation communities, my quantitative study supported the relevance of intrinsic motivation in community settings, but failed to confirm the relevance of extrinsic motivation.

Overall, the quantitative study presented in part IV confirmed most of the individual level propositions derived from qualitative analysis in part III. Yet, because individuals are influenced by aspects situated in their environments, two other levels of antecedents of social capital are relevant in corporate innovation communities. Related to that, the explained variance of the structural dimension of social capital remains fairly low in the SEM on individual level antecedents of social capital in corporate innovation communities. Hence, the findings with regard to the other two levels of antecedents of innovation community social capital (i.e., organizational and community level) constitute important contributions to the literature. I discuss these two remaining contributions in the following.

Third, empirical evidence presented in part III indicated a *relevance of organizational level antecedents* of social capital in corporate innovation community settings. As this dissertation focuses on communities that receive corporate support, aspects related to the supporting organization determine the environment in which the community operates to a large extent. This argument is especially tempting if a large part of community members is employed by the supporting organization. A practical example of such a community is the IBM Innovation Jam (Bjelland & Wood, 2008). With regard to organizational level antecedents, empirical data indicated a relevance of collaborative and innovative corporate culture as well as formal and informal organization as antecedents of community social capital. While a number of scholars have already argued for the relevance of cultural aspects (e.g., Leana & van Buren, 1999) and formal as well as informal organization (e.g., Brass et al., 2004) for corporate social capital in general, the research presented in this dissertation adds to the understanding of the role of these facets in community settings. A corporate culture that supports collaboration and innovation is positively related to the number of ties as well as shared visions and languages between community members. In addition, I propose that formal and informal structures determine network patterns of innovation community members who work for the supporting organization. An important contribution of my research is that community social capital builds upon organizational social capital. That is, community members convey and make use of existing network structures during innovation processes. For instance, if participants of a community workshop are able to draw on numerous existing network ties with other workshop participants they have developed in formal as well as informal settings beforehand, community members do not need to establish social capital before they are able to exchange knowledge successfully.

Fourth, this dissertation contributes to scientific knowledge on antecedents of innovation community social capital by highlighting *relevant facets on the community level*. For instance, data indicated that an appropriate level of creativity with respect to task design is positively related to shared visions and languages among community members. While the importance of task design in innovation processes has often been studied (Lakhani & Wolf, 2005; Müller-Seitz & Reger, 2009), I contribute to the understanding of this facet by conceptualizing task design as an antecedent of cognitive social capital in innovation community settings.

Moreover, data indicated that leadership in innovation communities affects cognitive social capital. While scholars have argued for the importance of leaders in community settings (Fleming & Waguespack, 2007), my framework contributes to the literature by clarifying the role of innovation community leaders with respect to shared visions. More specifically, empirical observations illustrate the role of community leaders to structure community activities without making use of hierarchical orders. As my interviewees have highlighted, such behavior is conducive to the development of a shared vision within the community.

Finally, the framework presented stresses the role of workshop as well as platform designs for the establishment of structural as well as cognitive social capital. I add to the literature as I emphasize the importance of design elements in both online as well as offline settings to strengthen interaction between community members as well as to support a shared vision and language within the community. While most of extant researchers have focused on communities that operate exclusively in offline settings (e.g., the windsurfing example provided by von Hippel (2001)) or on online platforms (e.g., the OSS example provided by Lerner and Tirole (2002) or Lakhani and von Hippel (2003)), the communities I have investigated make use of both interaction channels. Data indicated that offline interaction has a stronger influence on shared visions and trust than online interaction.

3 Managerial implications

The aim of my research is to help companies to establish and manage corporate innovation communities successfully to become more innovative. The framework offered in this dissertation suggests that community management is complex and calls for actions on operational as well as executive levels. Related to that, community management is not the task of a single position, but needs to be accomplished jointly by employees of different departments and hierarchical levels. Three groups of stakeholders are particularly important for successful community management: (1) T*he technology and innovation management department* that is responsible for the operational management of community work, which includes aspects such as task formulation, provision of workshop facilities, and online community platforms; (2) *the HR department*, which supports the technology and innovation management department in many ways, for instance in the recruiting of desired community members[46]; and (3) *the board of management*, which shapes the environment in which the corporate innovation community works, for instance with regard to corporate culture. In consequence, managerial implications discussed in this chapter address these three groups of corporate innovation community managers.

I structure the implications for community management in four categories: (1) Facets related to the self-reinforcing nature of social capital, (2) facets related to single community members and individual aspects, (3) facets related to the organizational environment of the community, and (4) facets related to the community as a collective. For each of these categories, I provide questions that help community managers to think about relevant aspects with regard to community network management. Furthermore, I provide underlying rationales from a managerial perspective why these questions are helpful and discuss performance indicators that allow community network management to assess current procedures and identify

[46] While the apportionment of work described mirrors a standard procedure, empirical observations presented in part III (multiple case study) show that different departments collaborate intensively in community management. In many cases, one department takes over tasks that would be assumed to be done by another department (e.g., the HR department organizes workshop facilities or provides an online platform).

potentials for improvement. Figure 27 summarizes the managerial implications of the research presented within this dissertation.

Managerial question	*Background*	*Performance indicators*
Do we spend sufficient resources for community support?	• Missing supporting activities hinder the successful development of community networks • This aspect is amplified by the interrelations of network structures, relationship quality, and shared visions	• Number of attendees at workshops or registrations on platform • Visibility of the community and emergence of specific terms • Frequency of workshops and traffic on platform
Do we recruit and motivate desired individuals for community participation?	• Personality traits influence the extent to which community members engage in networking	• Network centrality of community members in other settings (e.g., sports groups)
	• Motivations of members affect whether community work is perceived as a joint effort or individual task	• Reasons to engage in community work: Collaborative versus individual interests
Does our corporate environment support innovative efforts of community members?	• Corporate cultures associated with collaboration and innovation are beneficial for community networks	• Freedom of employees to engage in community activities • Willingness to help among employees
	• Formal and informal structures in the supporting organization provide interaction patterns for community work	• Boundary crossing processes and projects • Prevalence of opportunities for informal exchange
Do we support and enable working processes within the community appropriately?	• Creativity in task design has benefits and drawbacks	• Freedom of members to design their solution • Understanding what the supporting company aims to achieve
	• Community leaders guide the community on their way to the desired outcome	• Availability of a person (or group) who guides the community
	• Workshop and platform designs provide rooms in which community interaction takes place	• Appropriateness of facilities and materials provided • Functionalities and usability of platform

Figure 27: Summary of managerial implications

With regard to the first category, which relates to the self-reinforcing nature of social capital, community managers should ask themselves: *Do we spend sufficient resources for community support?* This question is highly relevant because missing supporting activities hinder the successful development of community networks, especially because of the interrelations between network structure, relationship quality, and shared visions within the community network. The research presented in this dissertation highlights that aspects such as trust, identification with the community as well as shared visions of members are both consequences of dense community networks as well as factors of influence for further interaction relationships.

More specifically, because of the interrelated nature of different dimensions of social capital, the successful management of one network dimension may be equalized by inappropriate activities with regard to another dimension. For instance, successful management of network structures may deflagrate if community management does not consider the establishment of a shared vision of community work. As an example, management support with regard to network structures via workshop and platform designs may fail to succeed if an inappropriate task design hinders the community to develop a shared vision or goal of community efforts. On the other hand, if for example a community leader succeeds in establishing a shared vision, trusted ties in the community will intensify, and interaction among innovation community members will flourish.

In a nutshell, because network structures, relationship quality (i.e., trust as well as frequency and closeness of interaction), and shared visions within the community are highly interrelated, community management needs to consider each social capital dimension in order to succeed. From an organizational perspective, a proactive approach to community management is pivotal. Related to the self-reinforcing nature of community networks and social capital, a holistic and integrated community management is able to produce remarkable benefits. In contrast, negligence is likely to lead to failure in establishing and managing innovation communities. Note that this implication stands in sharp contrast to the general assumption that managerial efforts in the context of informal community networks do not play a pivotal role.

Performance indicators to evaluate appropriate resource allocation to community network management touch all three dimensions of social capital. Community management is recommended to track developments of network structures, shared visions within the community, as well as the quality of the relationships in the community network. With regard to community network structures, growth rates of the community network in terms of new members or additional interaction relationships represent informative measures of success. In this context, log files of community platforms may be helpful. Management could analyze registrations of additional members or interaction patterns of existing community members. In addition, lists of attendees of community workshops can be studied. Developments of shared visions and languages are mirrored in the identification of community members with the community as a collective as well as with outputs of community innovation processes. To measure the performance with regard to identification, it is often useful to think about the visibility of the community as a collective. High identification of community members is in many cases accompanied by intense activities of community members to diffuse (preliminary) outcomes of community work in the organization. A shared language among community members is, obviously, reflected in the ease of interaction in innovation processes. To measure the performance in this context, log files of community platforms are again valuable. Here, a performance indicator is, for example, the emergence of specific terms the community agrees upon in order to step forward. Finally, with regard to quality of relationships within the community network, performance indicators for community management should address frequency of interaction, which could be measured via traffic on the community platform or the number of community workshops per month[47].

Second, the individual level category poses the following question to community management: *Do we recruit and motivate desired individuals for community participation?* This question is important because individual characteristics and motivations of community members affect successful development of community networks. With regard to individual characteristics, qualitative evidence suggested that all factors of the FFM are relevant, while the quantitative approach, applied in part IV, confirmed the role of extraversion, emotional stability, and conscientiousness

[47] Trust within the community network is hard to measure for community management because performance indicators in this context need to address the content of interaction among community members. Such analysis is in many cases restricted by data privacy. However, the research presented in this dissertation suggests that frequency and closeness of interaction affects trust significantly.

in the establishment of community networks. In this context, it is important to emphasize that my argument is not that these individuals contribute most important resources to community activities. However, their personality traits stimulate interaction and dense community networks as well as shared visions. Hence, while aspects such as knowledge and capabilities should of course be considered when selecting community members, personality influences the extent to which community members approach others within and develop identification towards the community. As a result, such candidates should be selected for community activities in order to foster social capital.

In addition, my findings suggest that community management should be aware of the consequences of incentives for community participation. While qualitative evidence presented suggests that both extrinsic and intrinsic motivations of community members are associated with successful network establishment, quantitative research failed to confirm the role of extrinsic rewards. These findings have important implications for the management of innovation communities. Some extrinsic rewards may also lead to rivalry between community members and, consequently, have detrimental effects on interaction and networks. To the contrary, intrinsic motivation is associated with a shared vision and language within the community, which leads to close, frequent, and trustful exchange. Hence, then deciding upon incentivizing community participation, community management should be aware that intrinsically motivated community members are likely to engage in networking activities in the community. To the contrary, extrinsic motivation is not necessarily related to intense interaction. In more detail, extrinsic motivation may be associated with rivalry between community members, for instance in the context of prizes in innovation contests. Rivalry often leads to reduced interaction as community members are afraid that exchanging ideas with others is disadvantageous for them. To the contrary, other extrinsic motivations, such as the desire to develop a reputation within the community, are beneficial for social capital as they are not strongly associated with rivalry. While rivalry may be beneficial for community outputs up to a certain threshold, community management needs to take the negative influence of rivalry on community networks into account as well.

A performance indicator that addresses issues with regard to recruiting and personality is the network centrality of community members in other settings, for instance within the supporting organization. As an example, community members who are particularly active within informal settings in the organization (e.g., sports groups) are precious because these people often possess valuable personality traits that are associated with networking. To monitor the appropriateness of incentivicing activities, community management is recommended to analyze the reasons why members engage in community activities. In this context, it is interesting to measure if community members participate because of collaborative interests or for individual success. If collaborative efforts are the dominant driver of community participation and motivations associated with individual success play a minor role, incentives are well aligned with the goal to establish a successful community network. An opportunity to measure motivations of members in this context is an online survey on the community platform.

The third category of managerial implications is concerned with relevant aspects on the level of the supporting organization. To address facets of this category, community management should answer the following question: *Does our corporate environment support innovative efforts of community members?* This question is important because the organizational environment in which the community operates affects the success of community network establishment to a significant extent. In this context, decisions at the executive level with regard to corporate culture as well as formal and informal organization are of particular importance. Concerning the first aspect, a culture associated with collaborative efforts and innovation is decisive for establishing networks characterized by joint cognitive visions in innovation communities. Of course, management is not able to change organizational culture within one day. But management should consider the crucial role of corporate culture in innovation communities and such cultural aspects can be influenced in a longer period of time. With regard to the second aspect, formal and informal organization and supporting activities (e.g., education courses or coffee rooms) affect corporate networks as these aspects provide interaction patterns upon which community social capital develops. Of course, management is not asked to change formal structures just to foster innovation community social capital. Yet, management should be mindful that organizational aspects are pivotal antecedents of innovation community social capital, especially in those cases where many community members are employed by the focal organization. In this context, arrangements that support informal networks over and above formal structures, for example sports groups, are often easily implemented.

Performance indicators for cultural aspects are freedom of employees to engage in community activities as well as willingness to help among employees in general. With regard to the freedom to engage in community work, different aspects can hinder employees to engage in community work. Hence, community management is recommended to take a look at the reasons why members cancel their participation in community processes. If the reasons behind non-attendance are related to workload in operational fields or instruction by supervisors to accomplish 'more important' things, there is evidence that the culture in which community members operate embraces efficiency over innovation. In such settings, community networks are hard to establish. With respect to informal and formal organization, performance indicators are the extent to which formal structures (i.e., processes and projects) are associated with collaboration across departmental and organizational boundaries. Such information can easily be gathered from organizational charts. Numerous formal boundary crossing ties provide a promising structure for community networks to develop. In addition, a crucial measure is the prevalence of informal exchange among employees and external stakeholders as well as corporate activities that support such behavior. These indicators can often be measured via interviews with employees of the organization that address the role of informal exchange. Furthermore, the prevalence of opportunities to exchange informally such as coffee rooms or sports groups represents an insightful measure for the relevance of informal networks.

Finally, to consider the fourth category and the community as a collective, the following question is helpful for community management: *Do we support and enable working processes within the community appropriately?* The findings presented highlight that design and configuration of innovation processes within the community affect the establishment of community networks to a significant extent. In this context, task design, community leadership, as well as workshop and platform designs influence the creation of social capital. Task design is crucial to support shared visions within innovation communities. Hence, management needs to consider both benefits and drawbacks of creativity in task designs with respect to cognitive social capital. For example, very open tasks that allow community members to decide creatively what kind of solution to develop are beneficial because once a community has agreed upon a solution, community members identify heavily with their work. In addition, community members keep in mind their resources and capabilities when deciding which kind of solution they set as their target outcome. However, creative and open tasks also have drawbacks with respect to cognitive social capital because community members struggle to identify the goal they are approaching at the beginning of community work. In consequence, companies are advised to take account of advantages and disadvantages of creative tasks and should choose a mediocre level of creativity. Numerous criteria allow companies to cut down too openly formulated tasks. For instance, community management can ask the community to focus on specific product groups. As another example, it may be useful to determine the configuration of the community output (e.g., process design or product concept) in advance.

Furthermore, community leaders play decisive roles in establishing shared visions and languages within community networks. Extant scholars have argued that these leaders emerge autonomously in innovation communities (Fleming & Waguespack, 2007). While this is obviously the case in independent innovation communities, management of the company that supports innovation communities could also assign responsibility for leadership tasks (e.g., summary of preliminary outputs) to one or more community members. If community management decides to do this, leadership in the community should not be associated with decisional authority towards other community members. The minimized role of hierarchy constitutes a major strength of community innovation processes.

Finally, the design of workshops and platforms affects community networks as well. Management is asked to make sure that interaction and shared visions between community members are supported during workshops. As suggested, there are numerous levers to foster exchange in workshop settings, e.g., introduction of participants, changing subgroups, rules for collaborative behavior or interactive methodologies, such as LEGO Serious Play. Community management is recommended to provide appropriate facilities and materials to the community in order to foster the creation of community social capital. In addition, the framework presented suggests that community platforms are able to complement and support interaction and shared visions within communities. In designing community platforms, this research points to the importance of user interfaces. Intuitive usability and implementation of corporate design are able to foster both structural as well as cognitive social capital. However, companies need to be aware that, in respect to social capital, interaction on online platforms is not as beneficial as interaction in face-to-face settings.

A performance indicator to evaluate task design is sufficient freedom of community members to define the outcome of the task to solve by themselves in combination with an understanding what the supporting organization aims to achieve with this task. Here, it is decisive that community members have the feeling that the focal company plans to implement (partial) results of the innovation process. With respect to community leadership, performance indicators are the availability of a person (or group) who helps to guide the community towards the desired output, for instance by structuring preliminary results or preparing community workshops. At the same time, a crucial performance indicator is the minimized role of hierarchical structures in community processes. A rule of thumb that is helpful in this context is that the community members with the highest hierarchical position in the supporting organization should not accomplish community leadership. Finally, indicators to evaluate performance with regard to community workshop and platform design address the immediate working environment the supporting organization provides for innovation processes. In this context, workshop facilities provided by the company need to be equipped with appropriate materials. Indicators for these aspects include both digital (e.g., computers, video projectors) as well as non-digital technologies (e.g., whiteboards, flipcharts). Community management should consult community leaders to determine which kind of equipment is needed. Often, community leaders have know-how with respect to a certain methodology that fosters interaction. Such methodologies demand in many cases specific materials that need to be organized by community management beforehand. Another performance indicator is the provision of community platforms with user interfaces that are characterized by high usability. Individuals who are not familiar with the focal community platform can best evaluate the usability from the perspective of potential members of the corporate innovation community. In addition, user interfaces should reflect the corporate identity of the supporting organization to foster shared visions among community members.

4 Limitations and avenues for future research

The results and contributions of the studies presented in this dissertation should be considered in the light of a number of limitations. While I have already discussed limitations and future research avenues that are specific to the qualitative and quantitative research approaches presented in part III and IV, respectively, I discuss in this chapter limitations that hold for both studies, thereby providing suggestions for future research efforts.

First, the studies presented in this dissertation focus on innovation communities that receive corporate support. Notwithstanding the relevance of supported innovation communities, the concept originates from settings where communities operate independently and self-organized (Franke & Shah, 2003; von Hippel, 2001). Many facets of the proposed framework on antecedents of social capital are specific to communities that receive corporate support. For instance, relevance of corporate culture as an antecedent of social capital in innovation communities is unique to supported communities. Hence, the factors that influence social capital in innovation communities which operate independently are likely to differ fundamently. While it is reasonable to argue that some of my findings are transferable to self-organized innovation communities (e.g., the role of personality as an antecedent of social capital), other aspects are specific to settings that involve organizational support (e.g., the role of formal organizational structures).

Furthermore, new businesses supported by IT, such as *innocentive.com*, often referred to as innovation intermediaries (Tran, Hsuan, & Mahnke, 2011), act as middlemen between companies and innovation communities. In such settings, communities do not receive support directly from the organization that is interested in outcomes of community innovation processes, but from an intermediary that offers community support as a service for companies. Again, antecedents of social capital might differ in such settings. For instance, it would be interesting to analyze if the corporate culture of the company that searches for a solution through the intermediary is still of importance for community social capital.

Second, most of the empirical data analyzed in this dissertation was gathered at early stages of innovation community activities. In essence, four of the five communities analyzed in part III as well as the community analysed in part IV did exist no longer than nine months. This aspect may limit the transferability of my findings to innovation communities that exist for a longer period. For instance, while many interviewees mentioned the role of introduction of participants in community workshops, this aspect is likely to be of minor importance in longer existing innovation communities, where community members have jointly participated in numerous workshops before. As another example, data indicated that knowledge complementarity did not affect the choice of interaction partners in innovation communities. This finding contradicts arguments of many previous studies on antecedents of network structures (e.g., Cowan & Jonard, 2009). I suggest that the relevance of knowledge complementarity as an antecedent of social capital may also change when communities exist longer because access to resources may be more important in later stages of innovation processes. In addition, community members are better equipped with information, which allows them to evaluate the resources that are accessible through other community members.

Third, there is need for future quantitative studies in the area of the research question approached within this dissertation. For instance, researchers could test the relationship between organizational culture as well as structures (organizational level antecedents) and corporate innovation community social capital. While requirements for empirical fields are demanding, scholars can make use of measures provided by extant publications in this context (e.g., Russell & Russel, 1992). In addition, researchers are asked to test relationships between community level antecedents (creative task design, leadership, as well as workshop and platform design) and social capital. Again, the literature provides useful measures that can be adopted (e.g., Sagie, Zaidman, Amichai-Hamburger, Te'eni, & Schwartz, 2002). Finally, it would be interesting to look at the interrelations between different levels of antecedents of community social capital on a quantitative basis. For example, does corporate culture affect leadership behavior in community settings? Does creative task design affect extrinsic and intrinsic motivations of individuals?

Fourth, the studies presented in this dissertation focus on antecedents of social capital and do not take consequences of social capital into account. In order to produce a comprehensive understanding of the role of social capital in innovation communities, I call for further studies that conceptualize social capital as a mediator between innovation community characteristics and performance. While first conceptual papers do exist (e.g., Rass, Dumbach, Danzinger, Bullinger, & Möslein, 2011), there is need for future empirical work that contributes to a holistic understanding of the role of social capital.

5 Conclusion

What are antecedents of social capital in corporate innovation communities? Motivated by numerous studies on the importance of innovation communities in general as well as the pivotal role of social relationships within these settings, I have examined this question using both qualitative as well as quantitative research methods. The research findings presented in this dissertation can be summarized in two major categories. First, social capital in innovation communities is self-reinforcing as multiple dimensions of social capital are related to each other in a circular manner. As this circle of relationships can result in a wheel of fortune (where social capital within the community reproduces itself), as well as in a doom loop (where social capital within the community destroys itself), antecedents of different dimensions of social capital are highly relevant for both the theoretical understanding of the rise of social capital as well as managerial efforts to influence social relationships between corporate innovation community members.

I have structured antecedents of interrelated dimensions of social capital on three interrelated levels, representing the second category of findings. On the individual level, which captures antecedents that are specific to individual community members, I have identified personality and motivations of community members as antecedents of social capital. On the organizational level, which captures antecedents that are related to the supporting organization and, thus, determine the environment in which the community enacts, data suggested that both formal and informal organization as well as corporate culture are important antecedents of social capital. Finally, the community level of antecedents entails aspects that affect the community as a collective and includes facets of creative task design, leadership as well as community workshop and platform design.

References

Acquaah, M. (2007). Managerial Social Capital, Strategic Orientation, and Organizational Performance in an Emerging Economy. *Strategic Management Journal, 28*(12), 1235-1255.

Adamczyk, S., Haller, J. B. A., Bullinger, A. C., & Möslein, K. M. (2011). Knowing is Silver, Listening is Gold: On the Importance and Impact of Feedback in IT-Based Innovation Contests. *Proceedings of the 10th International Conference on Business Informatics, Zurich* (pp. 221-230).

Adler, P. S., & Kwon, S.-W. (2002). Social Capital: Prospects for a New Concept. *Academy of Management Review, 27*(1), 17-40.

Agerfalk, P. J., & Fitzgerald, B. (2008). Outsourcing to an Unknown Workforce: Exploring Opensourcing as a Global Sourcing Strategy. *MIS Quarterly, 32*(2), 385-409.

Ahuja, G. (2000). Collaboration Networks, Structural Holes, and Innovation: A Longitudinal Study. *Administrative Science Quarterly, 45*(3), 425-455.

Allen, R. C. (1983). Collective Invention. *Journal of Economic Behavior and Organization, 4*(1), 1-24.

Almirall, E., & Casadesus-Masanell, R. (2010). Open Versus Closed Innovation: A Model of Discovery and Divergence. *Academy of Management Review, 35*(1), 27-47.

Alpert, M. I. (1972). Personality and the Determinants of Product Choice. *Journal of Marketing Research, 4*(2), 89-92.

Anand, N., Gardner, H. K., & Morris, T. (2007). Knowledge-Based Innovation: Emergence and Embedding of New Practice Areas in Management Consulting Firms. *Academy of Management Journal, 50*(2), 406-428.

Anderson, M. H. (2008). Social Networks and the Cognitive Motivation to Realize Network Opportunities: A Study of Managers' Information Gathering Behaviors. *Journal of Organizational Behavior, 29*(1), 51-78.

Armstrong, J. S., & Overton, T. S. (1977). Estimating Nonresponse Bias in Mail Surveys. *Journal of Marketing Research, 14*, 396-402.

Asakawa, K., Nakamura, H., & Sawada, N. (2010). Firms' Open Innovation Policies, Laboratories' External Collaborations, and Laboratories' R&D Performance. *R&D Management, 40*(2), 109-123.

Asendorpf, J. B., & Wilpers, S. (1998). Personality Effects on Social Relationships. *Journal of Personality and Social Psychology, 74*(6), 1531-1544.

Ashton, M. C. (1998). Personality and Job Performance: The Importance of Narrow Traits. *Journal of Organizational Behavior, 19*(3), 289-304.

Baba, Y., Shichijo, N., & Sedita, S. R. (2009). How Do Collaborations with Universities Affect Firms' Innovative Performance? The Role of "Pasteur Scientists" in the Advanced Materials Field. *Research Policy, 38*(5), 756-764.

Baba, Y., & Walsh, J. P. (2010). Embeddedness, Social Epistemology and Breakthrough Innovation: The Case of the Development of Statins. *Research Policy, 39*(4), 511-522.

Baburoglu, O. N., & Ravn, I. (1992). Normative Action Research. *Organization Studies, 13*(1), 19-34.

Backhaus, K., Erichson, E., Plinke, W., & Weiber, R. (2003). *Multivariate Analysemethoden*. Berlin: Springer.

Bae, J., Wezel, F. C., & Koo, J. (2011). Cross-Cutting Ties, Organizational Density, and New Firm Formation in the US Biotech Industry, 1994-98. *Academy of Management Journal, 54*(2), 295-311.

Bagozzi, R. P. (1981). Evaluating Structural Equation Models with Unobservable Variables and Measurement Error: A Comment. *Journal of Marketing Research, 18*, 375-381.

Bagozzi, R. P., & Dholakia, U. M. (2006). Open Source Software User Communities: A Study of Participation in Linux User Groups. *Management Science, 52*(7), 1099-1115.

Bagozzi, R. P., & Yi, Y. (1988). On the Evaluation of Structural Equation Models. *Journal of the Academy of Marketing Science, 16*(1), 74-94.

Bahemia, H., & Squire, B. (2010). A Contingent Perspective of Open Innovation in New Product Development Projects. *International Journal of Innovation Management, 14*(4), 603-627.

Baldwin, C., & von Hippel, E. (2011). Modeling a Paradigm Shift: From Producer Innovation to User and Open Collaborative Innovation. *Organization Science, 22*(6), 1399-1417.

Balka, K., Raasch, C., & Herstatt, C. (2010). How Open is Open Source? - Software and Beyond. *Creativity and Innovation Management, 19*(3), 248-256.

Balkundi, P., & Harrison, D. A. (2006). Ties, Leaders, and Time in Teams: Strong Inference About Network Structure's Effects on Team Viability and Performance. *Academy of Management Journal, 49*(1), 49-68.

Balkundi, P., & Kilduff, M. (2005). The Ties That Lead: A Social Network Approach to Leadership. *The Leadership Quarterly, 16*(6), 941-961.

Balkundi, P., Kilduff, M., Barsness, Z. I., & Michael, J. H. (2007). Demographic Antecedents and Performance Consequences of Structural Holes in Work Teams. *Journal of Organizational Behavior, 28*(2), 241-260.

Bansemir, B. (2012). *Organizational Innovation Communities*. Wiesbaden: Gabler.

Bansemir, B., Neyer, A.-K., & Möslein, K. M. (2012). Knowledge Exchange in Intra-Organizational Innovation Communities: The Role of Cognitive and Affective States. *Business Research, 5*(1), 1-16.

Barden, J. Q., & Mitchell, W. (2007). Disentangling the Influences of Leaders' Relational Embeddedness on Interorganizational Exchange. *Academy of Management Journal, 50*(6), 1440-1461.

Barrick, M. R., & Mount, M. K. (1991). The Big Five Personality Dimensions and Job Performance: A Meta-Analysis. *Personnel Psychology, 44*(1), 1-26.

Barrick, M. R., Mount, M. K., & Judge, T. A. (2001). Personality and Performance at the Beginning of the New Millennium: What Do We Know and Where Do We Go Next? *International Journal of Selection and Assessment, 9*(1/2), 9-30.

Batjargal, B., & Liu, M. (2004). Entrepreneurs' Access to Private Equity in China: The Role of Social Capital. *Organization Science, 15*(2), 159-172.

Baum, J. A. C., Cowan, R., & Jonard, N. (2010). Network-Independent Partner Selection and the Evolution of Innovation Networks. *Management Science, 56*(11), 2094-2110.

Bayus, B. L. (2012). Crowdsourcing New Product Ideas over Time: An Analysis of the Dell IdeaStorm Community. *Management Science*, 1-19.

Bearden, W. O., Sharma, S., & Teel, J. E. (1982). Sample Size Effects on Chi Square and Other Statistics Used in Evaluating Causal Models. *Journal of Marketing Research, 19*(4), 425-430.

Bechky, B. A. (2003). Sharing Meaning Across Occupational Communities: The Transformation of Understanding on a Production Floor. *Organization Science, 14*(3), 312-330.

Belliveau, M. A. (2005). Blind Ambition? The Effects of Social Networks and Institutional Sex Composition on the Job Search Outcomes of Elite Coeducational and Women's College Graduates. *Organization Science, 16*(2), 134-150.

Belliveau, M. A., O'Reilly III, C. A., & Wade, J. B. (1996). Social Capital at the Top: Effects of Social Similarity and Status on CEO Compensation. *Academy of Management Journal, 39*(6), 1568-1593.

Bentler, P. M. (1990). Comparative Fit Indexes in Structural Models. *Psychological Bulletin, 107*(2), 238-246.

Bentler, P. M., & Bonett, D. G. (1980). Significance Tests and Goodness of Fit in the Analysis of Covariance Structures. *Psychological Bulletin, 88*(3), 588-606.

Bercovitz, J., & Feldman, M. P. (2007). Fishing Upstream: Firm Innovation Strategy and University Research Alliances. *Research Policy, 36*(7), 930-948.

Berends, H., van Burg, E., & van Raaij, E. M. (2011). Contacts and Contracts: Cross-Level Network Dynamics in the Development of an Aircraft Material. *Organization Science, 22*(4), 940-960.

Bergenholtz, C. (2011). Knowledge Brokering: Spanning Technological and Network Boundaries. *European Journal of Innovation Management, 14*(1), 74-92.

Bhaskaran, S. R., & Krishnan, V. (2009). Effort, Revenue, and Cost Sharing Mechanisms for Collaborative New Product Development. *Management Science, 55*(7), 1152-1169.

Bian, Y. (1997). Bringing Strong Ties Back in: Indirect Ties, Network Bridges, and Job Searches in China. *American Sociological Review, 62*(3), 366-385.

Bitner, M. J., Brown, S. W., & Meuter, M. L. (2000). Technology Infusion in Service Encounters. *Journal of the Academy of Marketing Science, 28*(1), 138-149.

Bitzer, J., & Geishecker, I. (2010). Who Contributes Voluntarily to OSS? An Investigation Among German IT Employees. *Research Policy, 39*(1), 165-172.

Bjelland, O. M., & Wood, R. C. (2008). An Inside View of IBM's "Innovation Jam." *MIT Sloan Management Review, 50*(1), 32-40.

Björk, J., & Magnusson, M. (2009). Where Do Good Innovation Ideas Come from? Exploring the Influence of Network Connectivity on Innovation Idea Quality. *Journal of Product Innovation Management, 26*(6), 662-670.

Blatt, R. (2009). Tough Love: How Communal Schemas and Contracting Practices Build Relational Captial in Entrepreneurial Teams. *Academy of Management Review, 34*(3), 533-551.

Blyler, M., & Coff, R. W. (2003). Dynamic Capabilities, Social Capital, and Rent Appropriation: Ties That Split Pies. *Strategic Management Journal, 24*(7), 677-686.

Bogers, M. (2011). The Open Innovation Paradox: Knowledge Sharing and Protection in R&D Collaborations. *European Journal of Innovation Management, 14*(1), 93-117.

Bolino, M. C., Turnley, W. H., & Bloodgood, J. M. (2002). Citizenship Behavior and the Creation of Social Capital in Organizations. *Academy of Management Review, 27*(4), 505-522.

Bond, A. (2006). Appetite for Innovation. *Computing & Control Engineering Journal, 17*(1), 12-13.

Bono, J. E., & Anderson, M. H. (2005). The Advice and Influence Networks of Transformational Leaders. *Journal of Applied Psychology, 90*(6), 1306-1314.

Borst, I. (2010). *Understanding Crowdsourcing - Effects of Motivation and Rewards on Participation and Performance in Voluntary Online Activities*. Rotterdam: ERIM.

Bothner, M. S., Podolny, J. M., & Smith, E. B. (2011). Organizing Contests for Status: The Matthew Effect Versus the Mark Effect. *Management Science, 57*(3), 439-457.

Boudreau, K. J., Lacetera, N., & Lakhani, K. R. (2011). Incentives and Problem Uncertainty in Innovation Contests: An Empirical Analysis. *Management Science, 57*(5), 843-863.

Bourgeois, L. J., & Eisenhardt, K. M. (1988). Strategic Decision Processes in High Velocity Environments: Four Cases in the Microcomputer Industry. *Management Science, 34*(7), 816-835.

Bouty, I. (2000). Interpersonal and Interaction Influences on Informal Resource Exchanges Between R&D Researchers Across Organizational Boundaries. *Academy of Management Journal, 43*(1), 50-65.

Brass, D. J., Galaskiewicz, J., Greve, H. R., & Tsai, W. (2004). Taking Stock of Networks and Organizations: A Multilevel Perspective. *Academy of Management Journal, 47*(6), 795-817.

Brill, P. (2012). Toy Industry Outlook 2012. *Gifts & Decorative Accesories*, pp. 114-118.

Brockman, B. K., Rawlston, M. E., Jones, M. A., & Halstead, D. (2010). An Exploratory Model of Interpersonal Cohesiveness in New Product Development Teams. *Journal of Product Innovation Management, 27*(2), 201-219.

Brody, R. P., & Cunningham, S. M. (1968). Personality Variables and the Consumer Decision Process. *Journal of Marketing Research, 5*(1), 50-57.

Brown, J. S., & Duguid, P. (1991). Organizational Learning and Communities-of-Practice: Toward a Unified View of Working, Learning, and Innovation. *Organization Science, 2*(1), 40-57.

Brown, T. J., Mowen, J. C., Donavan, D. T., & Licata, J. W. (2002). The Customer Orientation of Service Workers: Personality Trait Effects on Self- and Supervisor Performance Ratings. *Journal of Marketing Research, 39*(1), 110-119.

Browne, M. W., & Cudeck, R. (1993). Alternative Ways of Assessing Model Fit. In K. A. Bollen & J. S. Long (Eds.), *Testing Structural Equation Models* (pp. 136-162). Thousand Oaks: Sage.

Bruneel, J., D'Este, P., & Salter, A. (2010). Investigating the Factors That Diminish the Barriers to University-Industry Collaboration. *Research Policy, 39*(7), 858-868.

Bullinger, A. C., Neyer, A.-K., Rass, M., & Möslein, K. M. (2010). Community-Based Innovation Contests: Where Competition Meets Cooperation. *Creativity and Innovation Management, 19*(3), 290-303.

Burt, R. S. (1992). *Structural Holes: The Social Structure of Competition.* Cambridge: Harvard University Press.

Burt, R. S. (1997). The Contingent Value of Social Capital. *Administrative Science Quarterly, 42*(2), 339-365.

Burt, R. S. (2004). Structural Holes and Good Ideas. *American Journal of Sociology, 110*(2), 349-399.

Burt, R. S. (2007). Secondhand Brokerage: Evidence on the Importance of Local Structure for Managers, Bankers, and Analysts. *Academy of Management Journal, 50*(1), 119-148.

Burt, R. S., Hogarth, R. M., & Michaud, C. (2000). The Social Capital of French and American Managers. Organization Science, 11(2), 123-147.

Camgoz, S. M., Karan, M. B., & Ergeneli, A. (2011). Relationship Between the Big-Five Personality and the Financial Performance of Fund Managers. *Current Topics in Management, 15*, 137-152.

Carbonell, P., & Rodriguez, A. I. (2006). Designing Teams for Speedy Product Development: The Moderating Effect of Technological Complexity. *Journal of Business Research, 59*(2), 225-232.

Carbonell, P., Rodríguez-Escudero, A. I., & Pujari, D. (2009). Customer Involvement in New Service Development: An Examination of Antecedents and Outcomes. *Journal of Product Innovation Management, 26*(5), 536-550.

Carpenter, M. A., Li, M., & Jiang, H. (2012). Social Network Research in Organizational Contexts: A Systematic Review of Methodological Issues and Choices. *Journal of Management, 38*(4), 1328-1361.

Carroll, G. R., & Teo, A. C. (1996). On the Social Networks of Managers. *Academy of Management Journal, 39*(2), 421-440.

Cattani, G., Ferriani, S., Negro, G., & Perretti, F. (2008). The Structure of Consensus: Network Ties, Legitimation, and Exit Rates of U.S. Feature Film Producer Organizations. *Administrative Science Quarterly, 53*(1), 145-182.

Chen, T. (2011). Personality Traits Hierarchy of Online Shoppers. *International Journal of Marketing Studies, 3*(4), 23-39.

Cherns, A. (1976). The Principles of Sociotechnical Design. *Human Relations, 29*(8), 783-792.

Chesbrough, H. W. (2003a). *Open Innovation: The New Imperative for Creating and Profiting from Technology*. Boston: Havard Buisness Press.
Chesbrough, H. W. (2003b). The Era of Open Innovation. *MIT Sloan Management Review, 44*(3), 35-41.
Chesbrough, H. W., & Crowther, A. K. (2006). Beyond High Tech: Early Adopters of Open Innovation in Other Industries. *R&D Management, 36*(3), 229-236.
Christensen, J. F., Olesen, M. H., & Kjar, J. S. (2005). The Industrial Dynamics of Open Innovation-Evidence from the Transformation of Consumer Electronics. *Research Policy, 34*(10), 1533-1549.
Chua, R. Y. J., Ingram, P., & Morris, M. W. (2008). From the Head and the Heart: Locating Cognition- and Affect-Based Trust in Managers' Professional Networks. *Academy of Management Journal, 51*(3), 436-452.
Chung, S. A., Singh, H., & Lee, K. (2000). Complementarity, Status Similarity and Social Capital as Drivers of Alliance Formation. *Strategic Management Journal, 21*(1), 1-22.
Chung, W., & Ng, T. W. (2011). Strategies of the Toy Manufacturers in the Pearl River Delta Area: From OEM to OBM. *Interdisciplinary Journal of Contemporary Research in Business, 3*(3), 50-64.
Cillo, P., de Luca, L. M., & Troilo, G. (2010). Market Information Approaches, Product Innovativeness, and Firm Performance: An Empirical Study in the Fashion Industry. *Research Policy, 39*(9), 1242-1252.
Coleman, J. S. (1988). Social Capital in the Creation of Human Capital. *American Journal of Sociology, 94*, 95-120.
Coleman, J. S. (1990). *Foundations of Social Theory*. Cambridge: Harvard University Press.
Collins, C. J., & Smith, K. G. (2006). Knowledge Exchange and Combination: The Role of Human Resource Practices in the Performance of High-Technology Firms. *Academy of Management Journal, 49*(3), 544-560.
Constant, D., Sproull, L., & Kiesler, S. (1996). The Kindness of Strangers: The Usefulness of Electronic Weak Ties for Technical Advice. *Organization Science, 7*(2), 119-135.
Cooper, R. G. (2001). *Winning at New Products - Accelerating the Process from Idea to Launch* (3rd ed.). New York: Basic Books.
Costa, P. T., & McCrae, R. R. (1992). *Revised NEO Personality Inventory (NEO PI-R) and NEO Five-Factor Inventory (NEO-FFI) Professional Manual*. Odessa: Psychological Assessment Resources.
Cotton, R. D., Shen, Y., & Livne-Tarandach, R. (2011). On Becoming Extraordinary: The Content and Structure of the Developmental Networks of Major League Baseball Hall of Famers. *Academy of Management Journal, 54*(1), 15-46.
Cowan, R., & Jonard, N. (2009). Knowledge Portfolios and the Organization of Innovation Networks. *Academy of Management Review, 34*(2), 320-342.
Cronbach, L. J. (1951). Coefficient Alpha and the Internal Structure of Tests. *Psychometrika, 16*(3), 297-334.

Cross, R., & Sproull, L. (2004). More Than an Answer: Information Relationships for Actionable Knowledge. *Organization Science, 15*(4), 446-462.

Dahlander, L., & Frederiksen, L. (2012). The Core and Cosmopolitans: A Relational View of Innovation in User Communities. *Organization Science, 23*(4), 988-1007.

Dahlander, L., & Gann, D. (2010). How Open is Innovation? *Research Policy, 39*(6), 699-709.

Dahlander, L., & Wallin, M. W. (2006). A Man on the Inside: Unlocking Communities as Complementary Assets. *Research Policy, 35*(8), 1243-1259.

Danzinger, F. (2010). *Kundeninteraktionskompetenz in Industriegütermärkten – Eine empirische Studie zur Interaktions- und Lernorientierung.* Wiesbaden: Gabler.

Datamonitor. (2010). Toys & Games in Germany.

Davis-Blake, A., & Uzzi, B. (1993). Determinants of Employment Externalization: A Study of Temporary Workers and Independent Contractors. *Administrative Science Quarterly, 38*(2), 195-223.

Deci, E. L. (1972). Intrinsic Motivation, Extrinsic Reinforcement, and Inequity. *Journal of Personality and Social Psychology, 22*(1), 113-120.

Dess, G. G., & Shaw, J. D. (2001). Voluntary Turnover, Social Capital, and Organizational Performance. *Academy of Management Review, 26*(3), 446-456.

Devaraj, S., Easley, R. F., & Crant, J. M. (2008). How Does Personality Matter? Relating the Five-Factor Model to Technology Acceptance and Use. *Information Systems Research, 19*(1), 93-105.

Diamantopoulos, A., & Siguaw, J. A. (2000). *Introducing Lisrel - A Guide for the Uninitiated.* London: Sage.

Digman, J. M. (1990). Personality Structure: Emergence of the Five-Factor Model. *Annual Review of Psychology, 41*(1), 417-440.

Dodgson, M., Gann, D., & Salter, A. (2006). The Role of Technology in the Shift Towards Open Innovation: The Case of Procter & Gamble. *R&D Management, 36*(3), 333-346.

Dokko, G., & Rosenkopf, L. (2010). Social Capital for Hire? Mobility of Technical Professionals and Firm Influence in Wireless Standards Committees. *Organization Science, 21*(3), 677-695.

Doll, B. (2009). *Prototyping zur Unterstützung sozialer Interaktionsprozesse.* Wiesbaden: Gabler.

Droge, C., Calantone, R., & Harmancioglu, N. (2008). New Product Success: Is It Really Controllable by Managers in Highly Turbulent Environments? *Journal of Product Innovation Management, 25*(3), 272-286.

Dyer Jr., W. G., & Wilkins, A. L. (1991). Better Stories, Not Better Constructs, to Generate Better Theory: A Rejoinder to Eisenhardt. *Academy of Management Review, 16*(3), 613-619.

Edwards, A. L. (1954). *Edwards Personal Preference Schedule.* Oxford: Psychological Corporation.

Eisenhardt, K. M. (1989). Building Theories from Case Study Research. *Academy of Management Review, 14*(4), 532-550.

Eisingerich, A. B., Bell, S. J., & Tracey, P. (2010). How Can Clusters Sustain Performance? The Role of Network Strength, Network Openness, and Environmental Uncertainty. *Research Policy, 39*(2), 239-253.

Enkel, E., Gassmann, O., & Chesbrough, H. W. (2009). Open R&D and Open Innovation: Exploring the Phenomenon. *R&D Management, 39*(4), 311-316.

Faems, D., Janssens, M., & van Looy, B. (2010). Managing the Co-Operation-Competition Dilemma in R&D Alliances: A Multiple Case Study in the Advanced Materials Industry. *Creativity and Innovation Management, 19*(1), 3-22.

Fang, E. (2011). The Effect of Strategic Alliance Knowledge Complementarity on New Product Innovativeness in China. *Organization Science, 22*(1), 158-172.

Faraj, S., Jarvenpaa, S. L., & Majchrzak, A. (2011). Knowledge Collaboration in Online Communities. *Organization Science, 22*(5), 1224-1239.

Faraj, S., & Johnson, S. L. (2011). Network Exchange Patterns in Online Communities. *Organization Science, 22*(6), 1464-1480.

Ferriani, S., Cattani, G., & Baden-Fuller, C. (2009). The Relational Antecedents of Project-Entrepreneurship: Network Centrality, Team Composition and Project Performance. *Research Policy, 38*(10), 1545-1558.

Fey, C. F., & Birkinshaw, J. (2005). External Sources of Knowledge, Governance Mode, and R&D Performance. *Journal of Management, 31*(4), 597-621.

Fischer, H. M., & Pollock, T. G. (2004). Effects of Social Capital and Power on Surviving Transformational Change: The Case of Initial Public Offerings. *Academy of Management Journal, 47*(4), 463-481.

Fleming, L., & Waguespack, D. M. (2007). Brokerage, Boundary Spanning, and Leadership in Open Innovation Communities. *Organization Science, 18*(2), 165-180.

Florin, J., Lubatkin, M., & Schulze, W. (2003). A Social Capital Model of High-Growth Ventures. *Academy of Management Journal, 46*(3), 374-384.

Fornell, C., & Larcker, D. F. (1981). Evaluating Structural Equation Models with Unobservable Variables and Measurement Error. *Journal of Marketing Research, 18*, 39-50.

Forret, M. L., & Dougherty, T. W. (2001). Correlates of Networking Behavior for Managerial and Professional Employees. *Group & Organization Management, 26*(3), 283-311.

Foss, N. J., Laursen, K., & Pedersen, T. (2010). Linking Customer Interaction and Innovation: The Mediating Role of New Organizational Practices. *Organization Science, 22*(4), 980-999.

Franke, N., Keinz, P., & Schreier, M. (2008). Complementing Mass Customization Toolkits with User Communities: How Peer Input Improves Customer Self-Design. *Journal of Product Innovation Management, 25*(6), 546-559.

Franke, N., & Shah, S. (2003). How Communities Support Innovative Activities: An Exploration of Assistance and Sharing Among End-Users. *Research Policy, 32*(1), 157-178.

Frey, B. S., & Oberholzer-Gee, F. (1997). The Cost of Price Incentives: An Empirical Analysis of Motivation Crowding-Out. *The American Economic Review, 87*(4), 746-755.

Frey, K., & Lüthje, C. (2011). Antecedents and Consequences of Interaction Quality in Virtual End-User Communities. *Creativity and Innovation Management, 20*(1), 22-35.

Füller, J. (2006). Why Consumers Engage in Virtual New Product Developments Initiated by Producers. *Advances in Consumer Research, 33*(1), 639-646.

Füller, J., Matzler, K., & Hoppe, M. (2008). Brand Community Members as a Source of Innovation. *Journal of Product Innovation Management, 25*(6), 608-619.

Gagné, M., & Deci, E. L. (2005). Self-Determination Theory and Work Motivation. *Journal of Organizational Behavior, 26*(4), 331-362.

Galaskiewicz, J., Bielefeld, W., & Dowell, M. (2006). Networks and Organizational Growth: A Study of Community Based Nonprofits. *Administrative Science Quarterly, 51*(3), 337-380.

Gargiulo, M., & Benassi, M. (2000). Trapped in Your Own Net? Network Cohesion, Structural Holes, and the Adaptation of Social Capital. *Organization Science, 11*(2), 183-196.

Gargiulo, M., Ertug, G., & Galunic, C. (2009). The Two Faces of Control: Network Closure and Individual Performance Among Knowledge Workers. *Administrative Science Quarterly, 54*(2), 299-333.

Garud, R., Dunbar, R. L. M., & Bartel, C. A. (2011). Dealing with Unusual Experiences: A Narrative Perspective on Organizational Learning. *Organization Science, 22*(3), 587-601.

Gassmann, O., & Enkel, E. (2004). Towards a Theory of Open Innovation: Three Core Process Archetypes. *R&D Management Conference* (pp. 1-18).

Gefen, D., Rigdon, E. E., & Straub, D. (2011). An Update and Extension to SEM Guidelines for Administrative and Social Science Research. *MIS Quarterly, 35*(2), 3-14.

Gilsing, V., Nooteboom, B., Vanhaverbeke, W., Duysters, G., & Vandenoord, A. (2008). Network Embeddedness and the Exploration of Novel Technologies: Technological Distance, Betweenness Centrality and Density. *Research Policy, 37*(10), 1717-1731.

Gist, M. E., & Mitchell, T. R. (1992). Self-Efficacy: A Theoretical Analysis of Its Determinants and Malleability. *Academy of Management Review, 17*(2), 183-211.

Gittelman, M. (2007). Does Geography Matter for Science-Based Firms? Epistemic Communities and the Geography of Research and Patenting in Biotechnology. *Organization Science, 18*(4), 724-741.

Glaser, B. C., & Strauss, A. L. (1967). *The Discovery of Grounded Theory: Strategies for Qualitative Research.* New York: Aldine.

Gnyawali, D. R., & Park, B.-J. (2011). Co-Opetition Between Giants: Collaboration with Competitors for Technological Innovation. *Research Policy, 40*(5), 650-663.

Goldberg, L. R. (1992). The Development of Markers for the Big-Five Factor Structure. *Psychological Assessment, 4*(1), 26-42.

Goldberg, L. R. (1993). The Structure of Phenotypic Personality Traits. *American Psychologist, 48*(1), 26-34.

Golden, B. R. (1992). The Past is the Past- Or is it? The Use of Retrospective Accounts as Indicators of Past Strategy. *Academy of Management Journal, 35*(4), 848-860.

Gosling, S. D., Rentfrow, P. J., & Swann Jr., W. B. (2003). A Very Brief Measure of the Big-Five Personality Domains. *Journal of Research in Personality, 37*(6), 504-528.

Granovetter, M. S. (1973). The Strength of Weak Ties. *American Journal of Sociology, 78*(6), 1360-1380.

Granovetter, M. S. (1985). Economic Action and Social Structure: The Problem of Embeddedness. *American Journal of Sociology, 91*(3), 481-510.

Grant, A. M. (2008). Does Intrinsic Motivation Fuel the Prosocial Fire? Motivational Synergy in Predicting Persistence, Performance, and Productivity. *Journal of Applied Psychology, 93*(1), 48-58.

Grant, A. M., & Berry, J. W. (2011). The Necessity of Others is the Mother of Invention: Intrinsic and Prosocial Motivations, Perspective-Taking, and Creativity. *Academy of Management Journal, 54*(1), 73-96.

Greer, C. R., & Lei, D. (2012). Collaborative Innovation with Customers: A Review of the Literature and Suggestions for Future Research. *International Journal of Management Reviews, 14*(1), 63-84.

Grewal, R., Lilien, G. L., & Mallapragada, G. (2006). Location, Location, Location: How Network Embeddedness Affects Project Success in Open Source Systems. *Management Science, 52*(7), 1043-1056.

Grindley, P. C., & Teece, D. J. (1997). Managing Intellectual Capital: Licensing and Cross-Licensing in Semiconductors and Electronics. *California Management Review, 39*(2), 8-41.

Grossman, E. B., Yli-Renko, H., & Janakiraman, R. (2012). Resource Search, Interpersonal Similarity, and Network Tie Valuation in Nascent Entrepreneurs' Emerging Networks. *Journal of Management, 38*(6), 1760-1787.

Gulati, R., & Puranam, P. (2009). Renewal Through Reorganization: The Value of Inconsistencies Between Formal and Informal Organization. *Organization Science, 20*(2), 422-440.

Haeussler, C. (2011). Information-Sharing in Academia and the Industry: A Comparative Study. *Research Policy, 40*(1), 105-122.

Hagedoorn, J. (1993). Understanding the Rationale of Strategic Technology Partnering: Interorganizational Modes of Cooperation and Sectoral Differences. *Strategic Management Journal, 14*(5), 371-385.

Hagedoorn, J., & Schakenraad, J. (1994). The Effect of Strategic Technology Alliances on Company Performance. *Strategic Management Journal, 15*(4), 291-309.

Hair, J. F., Black, W. C., Babin, B. J., & Anderson, R. E. (2010). *Multivariate Data Analysis - A Global Perspective*. New Jersey: Pearson Education.

Hansen, M. T. (1999). The Search-Transfer Problem: The Role of Weak Ties in Sharing Knowledge Across Organization Subunits. *Administrative Science Quarterly, 44*(1), 82-111.

Hansen, M. T. (2002). Knowledge Networks: Explaining Effective Knowledge Sharing in Multiunit Companies. *Organization Science, 13*(3), 232-248.

Hansen, M. T., Mors, M. L., & Løvås, B. (2005). Knowledge Sharing in Organizations: Multiple Networks, Multiple Phases. *Academy of Management Journal, 48*(5), 776-793.

Harary, F., & Batell, M. F. (1981). What is a System? *Social Networks, 3*, 29-40.

Harrigan, K. R. (1988). Joint Ventures and Competitive Strategy. *Strategic Management Journal, 9*(2), 141-158.

Harris, S. G., & Sutton, R. I. (1986). Functions of Parting Ceremonies in Dying Organizations. *Academy of Management Journal, 29*(1), 5-30.

Harrison, D., & Waluszewski, A. (2008). The Development of a User Network as a Way to Re-Launch an Unwanted Product. *Research Policy, 37*(1), 115-130.

Herstatt, C., & von Hippel, E. (1992). From Experience: Developing New Product Concepts Via the Lead User Method: A Case Study in a "Low-Tech" Field. *Journal of Product Innovation Management, 9*, 213-221.

Hertel, G., Niedner, S., & Herrmann, S. (2003). Motivation of Software Developers in Open Source Projects: An Internet-Based Survey of Contributors to the Linux Kernel. *Research Policy, 32*(7), 1159-1177.

Hienerth, C. (2006). The Commercialization of User Innovations: The Development of the Rodeo Kayak Industry. *R&D Management, 36*(3), 273-294.

Hislop, D. (2003). The Complex Relations Between Communities of Practice and the Implementation of Technological Innovations. *International Journal of Innovation Management, 7*(2), 163-188.

Hite, J. M., & Hesterly, W. S. (2001). The Evolution of Firm Networks: From Emergence to Early Growth of the Firm. *Strategic Management Journal, 22*(3), 275-286.

Hitt, M. A., Hoskisson, R. E., & Kim, H. (1997). International Diversification: Effects on Innovation and Firm Performance in Product-Diversified Firms. *Academy of Management Journal, 40*(4), 767-798.

Hofmans, J., Kuppens, P., & Allik, J. (2008). Is Short in Length Short in Content? An Examination of the Domain Representation of the Ten Item Personality Inventory Scales in Dutch Language. *Personality and Individual Differences, 45*(8), 750-755.

Holland, A. S., & Roisman, G. I. (2008). Big Five Personality Traits and Relationship Quality: Self-Reported, Observational, and Physiological Evidence. *Journal of Social and Personal Relationships, 25*(5), 811-829.

Hough, L. M. (1992). The "Big Five" Personality Variables-Construct Confusion: Description Versus Prediction. *Human Performance, 5*(1-2), 139-155.

Hu, L., & Bentler, P. M. (1998). Fit Indices in Covariance Structure Modeling: Sensitivity to Underparameterized Model Misspecification. *Psychological Methods, 3*(4), 424-453.

Hu, L., & Bentler, P. M. (1999). Cutoff Criteria for Fit Indexes in Covariance Structure Analysis: Conventional Criteria Versus New Alternatives. *Structural Equation Modeling, 6*(1), 1-55.

Huff, A. S., & Möslein, K. M. (2004). An Agenda for Understanding Individual Leadership in Corporate Leadership Systems. In C. L. Cooper (Ed.), *Leadership and Management in the 21st Century - Business Challenges of the Future* (pp. 248-270). Oxford: Oxford University Press.

Hughes, M., & Perrons, R. K. (2011). Shaping and Re-Shaping Social Capital in Buyer–Supplier Relationships. *Journal of Business Research, 64*(2), 164-171.

Hutter, K., Hautz, J., Füller, J., Mueller, J., & Matzler, K. (2011). Communitition: The Tension Between Competition and Collaboration in Community-Based Design Contests. *Creativity and Innovation Management, 20*(1), 3-21.

Ibarra, H., Kilduff, M., & Tsai, W. (2005). Zooming In and Out: Connecting Individuals and Collectivities at the Frontiers of Organizational Network Research. *Organization Science, 16*(4), 359-371.

Ili, S., Albers, A., & Miller, S. (2010). Open Innovation in the Automotive Industry. *R&D Management, 40*(3), 246-255.

Inkpen, A. C., & Tsang, E. W. K. (2005). Social Capital, Networks, and Knowledge Transfer. *Academy of Management Review, 30*(1), 146-165.

Ireland, R. D., Hitt, M. A., & Vaidyanath, D. (2002). Alliance Management as a Source of Competitive Advantage. *Journal of Management, 28*(3), 413-446.

Jack, S. L. (2005). The Role, Use and Activation of Strong and Weak Network Ties: A Qualitative Analysis. *Journal of Management Studies, 42*(6), 1233–1259.

James, E. H. (2000). Race-Related Differences in Promotions and Support: Underlying Effects of Human and Social Capital. *Organization Science, 11*(5), 493-508.

Janzik, L. (2010). Contribution and Participation in Innovation Communities: A Classification of Incentives and Motives. *International Journal of Innovation and Technology Management, 7*(3), 247-262.

Jeppesen, L. B., & Lakhani, K. R. (2010). Marginality and Problem-Solving Effectiveness in Broadcast Search. *Organization Science, 21*(5), 1016-1033.

Jeppesen, L. B., & Laursen, K. (2009). The Role of Lead Users in Knowledge Sharing. *Research Policy, 38*(10), 1582-1589.

Jick, T. D. (1979). Mixing Qualitative and Quantitative Methods: Triangulation in Action. *Administrative Science Quarterly, 24*(4), 602-611.

Judge, T. A., Heller, D., & Mount, M. K. (2002). Five-Factor Model of Personality and Job Satisfaction: A Meta-Analysis. *Journal of Applied Psychology, 87*(3), 530-541.

Judge, T. A., Higgins, C. A., Thoresen, C. J., & Barrick, M. R. (1999). The Big Five Personality Traits, General Mental Ability, and Career Success Across the Life Span. *Personnel Psychology, 52*(3), 621-652.

Judge, W. Q., Fryxell, G. E., & Dooley, R. S. (1997). The New Task of R&D Management: Creating Goal-Directed Communities for Innovation. *California Management Review, 39*(3), 72-85.

Jung, J. H., Schneider, C., & Valacich, J. (2010). Enhancing the Motivational Affordance of Information Systems: The Effects of Real-Time Performance Feedback and Goal Setting in Group Collaboration Environments. *Management Science, 56*(4), 724-742.

Jöreskog, K. G., & Sörbom, D. (1993). *Lisrel 8: Structural Equation Modeling with the Simplis Command Language*. Lincolnwood: Scientific Software International.

Kalnins, A., & Chung, W. (2006). Social Capital, Geography, and Survival: Gujarati Immigrant Entrepreneurs in the US Lodging Industry. *Management Science, 52*(2), 233-247.

Kehr, H. M. (2004). Integrating Implicit Motives, Explicit Motives, and Perceived Abilities: The Compensatory Model of Work Motivation and Volition. *Academy of Management Review, 29*(3), 479-499.

Kelley, M. R. (1994). Productivity and Information Technology: The Elusive Connection. *Management Science, 40*(11), 1406-1425.

Kelley, T. (2006). *The Ten Faces of Innovation - Strategies for Heightening Creativity*. London: Profile Books.

Kim, Y.-G., Lee, J.-N., Zmund, R. W., & Bock, G.-W. (2005). Behavioral Intention Formation in Knowledge Sharing: Examining the Roles of Extrinsic Motivators, Social-Psychological Forces, and Organizational Climate. *MIS Quarterly, 29*(1), 87-111.

Kirsch, L. J., Ko, D.-G., & Haney, M. H. (2010). Investigating the Antecedents of Team-Based Clan Control: Adding Social Capital as a Predictor. *Organization Science, 21*(2), 469-489.

Klein, K. J., Lim, B.-C., Saltz, J. L., & Mayer, D. M. (2004). How Do They Get There? An Examination of the Antecedents of Centrality in Team Networks. *Academy of Management Journal, 47*(6), 952-963.

Koka, B. R., & Prescott, J. E. (2002). Strategic Alliances as Social Capital: A Multidimensional View. *Strategic Management Journal, 23*(9), 795-816.

Kor, Y. Y., & Sundaramurthy, C. (2009). Experience-Based Human Capital and Social Capital of Outside Directors. *Journal of Management, 35*(4), 981-1006.

Kostova, T., & Roth, K. (2003). Social Capital in Multinational Corporations and a Micro-Macro Model of Its Formation. *Academy of Management Review, 28*(2), 297-317.

Kozlowski, S. W. J. (1987). Technological Innovation and Strategic HRM: Facing the Challenge of Change. *Human Resource Planning, 10*(2), 69-79.

Kraatz, M. S. (1998). Learning by Association? Interorganizational Networks and Adaptation to Environmental Change. *Academy of Management Journal, 41*(6), 621-643.

Krackhardt, D. (1992). The Strength of Strong Ties: The Importance of Philos in Organizations. In N. Nohria & R. G. Eccles (Eds.), *Networks and Organizations: Structure, Form and Action*. Boston, Massachusetts: Harvard Business School Press.

Kratzer, J., Gemünden, H. G., & Lettl, C. (2008). Revealing Dynamics and Consequences of Fit and Misfit Between Formal and Informal Networks in Multi-Institutional Product Development Collaborations. *Research Policy, 37*(8), 1356-1370.

Lakhani, K. R., & Wolf, R. G. (2005). Why Hackers Do What They Do: Understanding Motivation and Effort in Free/Open Source Software Projects. In J. Feller, B. Fitzgerald, S. Hissam, & K. R. Lakhani (Eds.), *Perspectives on Free and Open Source Software* (pp. 1-27). MIT Press.

Lakhani, K. R., & von Hippel, E. (2003). How Open Source Software Works: "Free" User-To-User Assistance. *Research Policy, 32*(6), 923-943.

Lange, D., Boivie, S., & Henderson, A. D. (2009). The Parenting Paradox: How Multibusiness Diversifiers Endorse Disruptive Technologies While Their Corporate Children Struggle. *Academy of Management Journal, 52*(1), 179-198.

Lau, A. K. W., Tang, E., & Yam, R. C. M. (2010). Effects of Supplier and Customer Integration on Product Innovation and Performance: Empirical Evidence in Hong Kong Manufacturers. *Journal of Product Innovation Management, 27*(5), 761-777.

Laursen, K., Masciarelli, F., & Prencipe, A. (2011). Regions Matter: How Localized Social Capital Affects Innovation and External Knowledge Acquisition. *Organization Science, 23*(1), 177-193.

Laursen, K., & Salter, A. (2006). Open for Innovation: The Role of Openness in Explaining Innovation Performance Among U.K. Manufacturing Firms. *Strategic Management Journal, 27*(2), 131-150.

Lavie, D., & Drori, I. (2012). Collaborating for Knowledge Creation and Application: The Case of Nanotechnology Research Programs. *Organization Science, 23*(3), 704-724.

Lawson, B., Petersen, K. J., Cousins, P. D., & Handfield, R. B. (2009). Knowledge Sharing in Interorganizational Product Development Teams: The Effect of Formal and Informal Socialization Mechanisms. *Journal of Product Innovation Management, 26*(2), 156-172.

Lawson, B., & Samson, D. (2001). Developing Innovation Capability in Organisations: A Dynamic Capabilities Approach. *International Journal of Innovation Management, 5*(3), 377-400.

Leana, C. R., & Pil, F. K. (2006). Social Capital and Organizational Performance: Evidence from Urban Public Schools. *Organization Science, 17*(3), 353-366.

Leana, C. R., & van Buren, H. J. (1999). Organizational Social Capital and Employment Practices. *Academy of Management Review, 24*(3), 538-555.

Lechner, C., Frankenberger, K., & Floyd, S. W. (2010). Task Contingencies in the Curvilinear Relationships Between Intergroup Networks and Initiative Performance. *Academy of Management Journal, 53*(4), 865-889.

Lee, J. (2010). Heterogeneity, Brokerage, and Innovative Performance: Endogenous Formation of Collaborative Inventor Networks. *Organization Science, 21*(4), 804-822.

Leimeister, J. M., Huber, M., Bretschneider, U., & Krcmar, H. (2009). Leveraging Crowdsourcing - Theory-Driven Design, Implementation and Evaluation of Activation-Supporting Components for IT-Based Idea Competitions. *Journal of Management Information Systems, 26*(1), 197-224.

Leiponen, A., & Helfat, C. E. (2010). Innovation Objectives, Knowledge Sources, and the Benefits of Breadth. *Strategic Management Journal, 31*, 224-236.

Lerner, J., & Tirole, J. (2002). Some Simple Economics of Open Source. *Journal of Industrial Economics, 50*(2), 197-234.

Lettl, C., Rost, K., & von Wartburg, I. (2009). Why Are Some Independent Inventors "Heroes" and Others "Hobbyists"? The Moderating Role of Technological Diversity and Specialization. *Research Policy, 38*(2), 243-254.

Levin, D. Z., & Cross, R. (2004). The Strength of Weak Ties You Can Trust: The Mediating Role of Trust in Effective Knowledge Transfer. *Management Science, 50*(11), 1477-1490.

Levin, D. Z., Walter, J., & Murnighan, J. K. (2011). Dormant Ties: The Value Of Reconnecting. *Organization Science, 22*(4), 923-939.

Lewin, K. (1947). Frontiers in Group Dynamics II. *Human Relations, 1*(2), 143-153.

Li, J. J., Poppo, L., & Zhou, K. Z. (2008). Do Managerial Ties in China Always Produce Value? Competition, Uncertainty, and Domestic Versus Foreign Firms. *Strategic Management Journal, 29*(4), 383-400.

Lilien, G. L., Morrison, P. D., Searls, K., Sonnack, M., & von Hippel, E. (2002). Performance Assessment of the Lead User Idea-Generation Process for New Product Development. *Management Science, 48*(8), 1042-1059.

Lin, H.-F. (2007). Effects of Extrinsic and Intrinsic Motivation on Employee Knowledge Sharing Intentions. *Journal of Information Science, 33*(2), 135-149.

Lin, N. (1999). Social Networks and Status Attainment. *Annual Review of Sociology, 25*(1), 467-487.

Lingo, E. L., & O'Mahony, S. (2010). Nexus Work: Brokerage on Creative Projects. *Administrative Science Quarterly, 55*(1), 47-81.

Lovas, B., & Ghoshal, S. (2000). Strategy as Guided Evolution. *Strategic Management Journal, 21*(9), 875-896.

Lovejoy, W. S., & Sinha, A. (2010). Efficient Structures for Innovative Social Networks. *Management Science, 56*(7), 1127-1145.

Lüthje, C., Herstatt, C., & von Hippel, E. (2005). User-Innovators and "Local" Information: The Case of Mountain Biking. *Research Policy, 34*(6), 951-965.

Ma, M., & Agarwal, R. (2007). Through a Glass Darkly: Information Technology Design, Identity Verification, and Knowledge Contribution in Online Communities. *Information Systems Research, 18*(1), 42-67.

Mahmood, I. P., Zhu, H., & Zajac, E. J. (2011). Where Can Capabilities Come From? Network Ties and Capability Acquisition in Business Groups. *Strategic Management Journal, 32*(8), 820-848.

Mahr, D., & Lievens, A. (2012). Virtual Lead User Communities: Drivers of Knowledge Creation for Innovation. *Research Policy, 41*(1), 167-177.

Malone, T. W., Laubacher, R., & Dellarocas, C. (2009). Harnessing Crowds: Mapping the Genome of Collective Intelligence. Cambridge.

Martin, J. A., & Eisenhardt, K. M. (2010). Rewiring: Cross-Business-Unit Collaborations in Multibusiness Organizations. *Academy of Management Journal, 53*(2), 265-301.

Matzler, K., Bidmon, S., & Grabner-Kräuter, S. (2006). Individual Determinants of Brand Affect: The Role of the Personality Traits of Extraversion and Openness to Experience. *Journal of Product & Brand Management, 15*(7), 427-434.

Maurer, I., Bartsch, V., & Ebers, M. (2011). The Value of Intra-Organizational Social Capital: How it Fosters Knowledge Transfer, Innovation Performance, and Growth. *Organization Studies, 32*(2), 157-185.

Maurer, I., & Ebers, M. (2006). Dynamics of Social Capital and Their Performance Implications: Lessons from Biotechnology Start-Ups. *Administrative Science Quarterly, 51*(2), 262-292.

McCrae, R. R., & Costa Jr., P. T. (1987). Validation of the Five-Factor Model of Personality Across Instruments and Observers. *Journal of Personality and Social Psychology, 52*(1), 81-90.

McCrae, R. R., & Costa Jr., P. T. (1989). Rotation to Maximize the Construct Validity of Factors in the NEO Personality Inventory. *Multivariate Behavioral Research, 24*(1), 107-124.

McCrae, R. R., & Costa, P. T. (1997). Personality Trait Structure as a Human Universal. *The American Psychologist, 52*(5), 509-516.

McDonald, R. P., & Ho, M.-H. R. (2002). Principles and Practice in Reporting Structural Equation Analyses. *Psychological Methods, 7*(1), 64-82.

McElroy, J. C., Hendrickson, A. R., Townsend, A. M., & DeMarie, S. M. (2007). Dispositional Factors in Internet Use: Personality Versus Cognitive Style. *MIS Quarterly, 31*(4), 809-820.

McEvily, B., Jaffee, J., & Tortoriello, M. (2012). Not All Bridging Ties Are Equal: Network Imprinting and Firm Growth in the Nashville Legal Industry, 1933-1978. *Organization Science, 23*(2), 547-563.

McFadyen, M. A., & Cannella Jr., A. A. (2004). Social Capital and Knowledge Creation: Diminishing Returns of the Number and Strength of Exchange Relationships. *Academy of Management Journal, 47*(5), 735-746.

McFadyen, M. A., Semadeni, M., & Cannella Jr., A. A. (2009). Value of Strong Ties to Disconnected Others: Examining Knowledge Creation in Biomedicine. *Organization Science, 20*(3), 552-564.

McLure Wasko, M., & Faraj, S. (2000). "It is What One Does": Why People Participate and Help Others in Electronic Communities of Practice. *Journal of Strategic Information Systems, 9*, 155-173.

McLure Wasko, M., & Faraj, S. (2005). Why Should I Share? Examining Social Capital and Knowledge Contribution in Electronic Networks of Practice. *MIS Quarterly, 29*(1), 35-57.

Moldoveanu, M. C., & Baum, J. A. C. (2011). "I Think You Think I Think You're Lying": The Interactive Epistemology of Trust in Social Networks. *Management Science, 57*(2), 393-412.

Molina-Morales, F. X., & Martínez-Fernández, M. T. (2009). Too Much Love in the Neighborhood Can Hurt: How an Excess of Intensity and Trust in Relationships May Produce Negative Effects on Firms. *Strategic Management Journal, 30*(9), 1013-1023.

Moliterno, T. P., & Mahony, D. M. (2011). Network Theory of Organization: A Mulitlevel Approach. *Journal of Management, 37*(2), 443-467.

Moran, P. (2005). Structural Versus Relational Embeddedness: Social Capital and Managerial Performance. *Strategic Management Journal, 26*(12), 1129-1151.

Mors, M. L. (2010). Innovation in a Global Consulting Firm: When the Problem is too Much Diversity. *Strategic Management Journal, 31*(8), 841-872.

Mosey, S., & Wright, M. (2007). From Human Capital to Social Capital: A Longitudinal Study of Technology-Based Academic Entrepreneurs. *Entrepreneurship Theory and Practice, 31*(6), 909-935.

Mount, M. K., Barrick, M. R., & Stewart, G. L. (1998). Five-Factor Model of Personality and Performance in Jobs Involving Interpersonal Interactions. *Human Performance, 11*(2/3), 145-165.

Mowen, J. C., & Spears, N. (1999). Understanding Compulsive Buying Among College Students: A Hierarchical Approach. *Journal of Consumer Psychology, 8*(4), 407-430.

Muck, P. M., Hell, B., & Gosling, S. D. (2007). Construct Validation of a Short Five-Factor Model Instrument. *European Journal of Psychological Assessment, 23*(3), 166-175.

Muhdi, L., & Boutellier, R. (2011). Motivational Factors Affecting Participation and Contribution of Members in Two Different Swiss Innovation Communities. *International Journal of Innovation Management, 15*(3), 543-562.

Müller-Seitz, G., & Reger, G. (2009). Is Open Source Software Living Up to its Promises? Insights for Open Innovation Management from Two Open Source Software-Inspired Projects. *R&D Management, 39*(4), 372-381.

Nahapiet, J., & Ghoshal, S. (1998). Social Capital, Intellectual Capital, and the Organizational Advantage. *Academy of Management Review, 23*(2), 242-266.

Nalebuff, B., & Ayres, I. (2004). Encouraging Suggestive Behavior. *Harvard Business Review*, p. 18.

Nambisan, S. (2002). Designing Virtual Customer Environments for New Product Development: Toward a Theory. *Academy of Management Review, 27*(3), 392-413.

Nambisan, S., & Baron, R. A. (2009). Virtual Customer Environments: Testing a Model of Voluntary Participation in Value Co-Creation Activities. *Journal of Product Innovation Management, 26*(4), 388-406.

Nambisan, S., & Baron, R. A. (2010). Different Roles, Different Strokes: Organizing Virtual Customer Environments to Promote Two Types of Customer Contributions. *Organization Science, 21*(2), 554-572.

Neyer, A.-K., Bullinger, A. C., & Möslein, K. M. (2009). Integrating Inside and Outside Innovators: A Sociotechnical Systems Perspective. *R&D Management, 39*(4), 410-419.

Nielsen, J. (2000). *Designing Web Usability*. Indianapolis: New Riders.

Norman, W. T. (1963). Toward an Adequate Taxonomy of Personality Attributes: Replicated Factor Structure in Peer Nomination Personality Ratings. *The Journal of Abnormal and Social Psychology, 66*(6), 574-583.

Nov, O. (2007). What Motivates Wikipedians? *Communications of the ACM, 50*(11), 60-64.

Obstfeld, D. (2005). Social Networks, the Tertius Iungens Orientation, and Involvement in Innovation. *Administrative Science Quarterly, 50*(1), 100-130.

Oh, H., Chung, M.-H., & Labianca, G. (2004). Group Social Capital and Group Effectiveness: The Role of Informal Socializing Ties. *Academy of Management Journal, 47*(6), 860-875.

Oh, H., & Kilduff, M. (2008). The Ripple Effect of Personality on Social Structure: Self-Monitoring Origins of Network Brokerage. *Journal of Applied Psychology, 93*(5), 1155-1164.

Oh, H., Labianca, G., & Chung, M.-H. (2006). A Multilevel Model of Group Social Capital. *Academy of Management Review, 31*(3), 569-582.

Oliveira, P., & von Hippel, E. (2011). Users as Service Innovators: The Case of Banking Services. *Research Policy, 40*(6), 806-818.

Ozcan, P., & Eisenhardt, K. M. (2009). Origin of Alliance Portfolios: Entrepreneurs, Network Strategies, and Firm Performance. *Academy of Management Journal, 52*(2), 246-279.

Packalen, K. A. (2007). Complementing Capital: The Role of Status, Demographic Features, and Social Capital in Founding Teams' Abilities to Obtain Resources. *Entrepreneurship Theory and Practice, 31*(6), 873-891.

Paruchuri, S. (2010). Intraorganizational Networks, Interorganizational Networks, and the Impact of Central Inventors: A Longitudinal Study of Pharmaceutical Firms. *Organization Science, 21*(1), 63-80.

Payne, G. T., Moore, C. B., Griffis, S. E., & Autry, C. W. (2011). Multilevel Challenges and Opportunities in Social Capital Research. *Journal of Management, 37*(2), 491-520.

Pennings, J. M., Lee, K., & van Witteloostuijn, A. (1998). Human Capital, Social Capital, and Firm Dissolution. *Academy of Management Journal, 41*(4), 425-440.

Perry-Smith, J. E. (2006). Social Yet Creative: The Role of Social Relationships in Facilitating Individual Creativity. *Academy of Management Journal, 49*(1), 85-101.

Perry-Smith, J. E., & Shalley, C. E. (2003). The Social Side of Creativity: A Static and Dynamic Social Network Perspective. *Academy of Management Review, 28*(1), 89-106.

Petersen, K. J., Handfield, R. B., & Ragatz, G. L. (2003). A Model of Supplier Integration into New Product Development. *Journal of Product Innovation Management, 20*(4), 284-299.

Peterson, R. A. (2000). A Meta-Analysis of Variance Accounted for and Factor Loadings in Exploratory Factor Analysis. *Marketing Letters, 11*(3), 261-275.

Pettigrew, A. M. (1990). Longitudinal Field Research on Change: Theory and Practice. *Organization Science, 1*(3), 267-292.

Phelps, C. C. (2010). A Longitudinal Study of the Influence of Alliance Network Structure and Composition on Firm Exploratory Innovation. *Academy of Management Journal, 53*(4), 890-913.

Phelps, C., Heidl, R., & Wadhwa, A. (2012). Knowledge, Networks, and Knowledge Networks: A Review and Research Agenda. *Journal of Management, 38*(4), 1115-1166.

Pil, F. K., & Leana, C. R. (2009). Applying Organizational Research to Public School Reform: The Effects of Teacher Human and Social Capital on Student Performance. *Academy of Management Journal, 52*(6), 1101-1124.

Piller, F. T., & Walcher, D. (2006). Toolkits for Idea Competitions: A Novel Method to Integrate Users in New Product Development. *R&D Management, 36*(3), 307-318.

Pintrich, P. R., Smith, D. A. F., Garcia, T., & Mckeachie, W. J. (1991). A Manual for the Use of the Motivated Strategies for Learning Questionnaire (MSLQ).

Pintrich, P. R., Smith, D. A. F., Garcia, T., & Mckeachie, W. J. (1993). Reliability and Predictive Validity of the Motivated Strategies for Learning Questionnaire (MSLQ). *Educational and Psychological Measurement, 53*(3), 801-813.

Pisano, G. P., & Verganti, R. (2008). Which Kind of Collaboration is Right for You? *Harvard Business Review, 86*(12), 78-86.

Podsakoff, P. M., & Organ, D. W. (1986). Self-Reports in Organizational Research: Problems and Prospects. *Journal of Management, 12*(4), 531-544.

Polidoro Jr., F., Ahuja, G., & Mitchell, W. (2011). When the Social Structure Overshadows Competitive Incentives: The Effects of Network Embeddedness on Joint Venture Dissolution. *Academy of Management Journal, 54*(1), 203-223.

Porter, M. E. (1991). The Competitive Advantage of Nations. In C. A. Montgomery & M. E. Porter (Eds.), *Strategy - Seeking and Securing Competitive Advantage*. Boston: Harvard Business Press.

Pruden, H. O., & Peterson, R. A. (1971). Personality and Performance-Satisfaction of Industrial Salesmen. *Journal of Marketing Research, 8*(11), 501-505.

Rass, M., Dumbach, M., Danzinger, F., Bullinger, A. C., & Möslein, K. M. (2011). Open Innovation and Firm Performance: Investigating the Mediating Role of Social Capital. *R&D Management Conference*. Norrköping, Sweden.

Reagans, R., & McEvily, B. (2003). Network Structure and Knowledge Transfer: The Effects of Cohesion and Range. *Administrative Science Quarterly, 48*(2), 240-267.

Roberts, J. A., Hann, I.-H., & Slaughter, S. A. (2006). Understanding the Motivations, Participation, and Performance of Open Source Software Developers: A Longitudinal Study of the Apache Projects. *Management Science, 52*(7), 984-999.

Rodan, S., & Galunic, C. (2004). More Than Network Structure: How Knowledge Heterogeneity Influences Managerial Performance and Innovativeness. *Strategic Management Journal, 25*(6), 541-562.

Rogers, E. M. (1995). *Diffusion of Innovations*. New York: Free Press.

Romero, E., Villar, P., Gómez-Fraguela, J. A., & López-Romero, L. (2012). Measuring Personality Traits with Ultra-Short Scales: A Study of the Ten Item Personality Inventory (TIPI) in a Spanish Sample. *Personality and Individual Differences, 53*(3), 289-293.

Rost, K. (2011). The Strength of Strong Ties in the Creation of Innovation. *Research Policy, 40*(4), 588-604.

Rothenberg, S., & Ettlie, J. E. (2011). Strategies to Cope with Regulatory Uncertainty in the Auto Industry. *California Management Review, 54*(1), 126-144.

Rowley, T., Behrens, D., & Krackhardt, D. (2000). Redundant Governance Structures: An Analysis of Structural and Relational Embeddedness in the Steel and Semiconductor Industries. *Strategic Management Journal, 21*(3), 369-386.

Russell, R. D., & Russel, C. J. (1992). An Examination of the Effects of Organizational Norms, Organizational Structure, and Environmental Uncertainty on Entrepreneurial Strategy. *Journal of Management, 18*(4), 639-656.

Sagie, A., Zaidman, N., Amichai-Hamburger, Y., Te'eni, D., & Schwartz, D. G. (2002). An Empirical Assessment of the Loose-Tight Leadership Model: Quantitative and Qualitative Analyses. *Journal of Organizational Behavior, 23*(3), 303-320.

Salter, A., & Martin, B. R. (2001). The Economic Benefits of Publicly Funded Basic Research: A Critical Review. *Research Policy, 30*, 509-532.

Sasovova, Z., Mehra, A., Borgatti, S. P., & Schippers, M. C. (2010). Network Churn: The Effects of Self-Monitoring Personality on Brokerage Dynamics. *Administrative Science Quarterly, 55*(4), 639-670.

Saucier, G. (1994). Mini-Markers: A Brief Version of Goldberg's Unipolar Big-Five Markers. *Journal of Personality Assessment, 63*(3), 506-516.

Sawhney, M., & Prandelli, E. (2000). Communities of Creation: Managing Distributed Innovation in Turbulent Markets. *California Management Review, 42*(4), 24-54.

Schermelleh-Engel, K., Moosbrugger, H., & Müller, H. (2003). Evaluating the Fit of Structural Equation Models: Tests of Significance and Descriptive Goodness-of-Fit Measures. *Methods of Psychological Research Online, 8*(2), 23-74.

Schnell, R., Hill, P. B., & Esser, E. (2011). *Methoden der empirischen Sozialforschung*. München: Oldenbourg Verlag.

Schrage, M. (2000). *Serious Play - How the World's Best Companies Simulate to Innovate*. Boston: Harvard Business School Press.

Schreier, M., & Prügl, R. (2008). Extending Lead-User Theory: Antecedents and Consequences of Consumers' Lead Userness. *Journal of Product Innovation Management, 25*(4), 331-346.

Schreyögg, G., & Sydow, J. (2010). Organizing for Fluidity? Dilemmas of New Organizational Forms. *Organization Science, 21*(6), 1251-1262.

Schröder, A., & Hölzle, K. (2010). Virtual Communities for Innovation: Influence Factors and Impact on Company Innovation. *Creativity and Innovation Management, 19*(3), 257-268.

Schultz, M., & Hatch, M. J. (2003). The Cycles of Corporate Branding: The Case of the LEGO Company. *California Management Review, 46*(1), 6-26.

Schumpeter, J. A. (1939). *Business Cycles: A Theoretical, Historical, and Statistical Analysis of the Capitalist Process*. New York: McGraw-Hill Book Company.

Schumpeter, J. A. (1942). *Capitalism, Socialism and Democracy*. New York: Harper.

Seibert, S. E., Kraimer, M. L., & Liden, R. C. (2001). A Social Capital Theory of Career Success. *Academy of Management Journal, 44*(2), 219-237.

Shaner, J., & Maznevski, M. (2011). The Relationship Between Networks, Institutional Development, and Performance in Foreign Investments. *Strategic Management Journal, 32*(5), 556-568.

Shaw, J. D., Duffy, M. K., Johnson, J. L., & Lockhart, D. E. (2005). Turnover, Social Capital Losses, and Performance. *Academy of Management Journal, 48*(4), 594-606.

Shipilov, A. V. (2009). Firm Scope Experience, Historic Multimarket Contact with Partners, Centrality, and the Relationship Between Structural Holes and Performance. *Organization Science, 20*(1), 85-106.

Smith, A. (1976). *An Inquiry into the Nature and Causes of the Wealth of Nations. Readings in Economic Sociology*. Chicago: The University of Chicago Press.

Somaya, D., Williamson, I. O., & Lorinkova, N. (2008). Gone But Not Lost: The Different Performance Impacts of Employee Mobility Between Cooperators Versus Competitors. *Academy of Management Journal, 51*(5), 936-953.

Sosa, M. E. (2011). Where Do Creative Interactions Come from? The Role of Tie Content and Social Networks. *Organization Science, 22*(1), 1-21.

Stam, W. (2009). When Does Community Participation Enhance the Performance of Open Source Software Companies? *Research Policy, 38*(8), 1288-1299.

Stam, W., & Elfring, T. (2008). Entrepreneurial Orientation and New Venture Performance: The Moderating Role of Intra- and Extraindustry Social Capital. *Academy of Management Journal, 51*(1), 97-111.

Steiger, J. H., & Lind, J. C. (1980). Statistically Based Tests for the Number of Common Factors. *Annual Meeting of the Psychometric Society*. Iowa City, IA.

Studt, T. (2003). 3M - Where Innovation Rules. *R&D Magazine*, 20-23.

Subramaniam, M., & Youndt, M. A. (2005). The Influence of Intellectual Capital on the Types of Innovative Capabilities. *Academy of Management Journal, 48*(3), 450-463.

Takeishi, A. (2001). Bridging Inter- and Intra-Firm Boundaries: Management of Supplier Involvement in Automobile Product Development. *Strategic Management Journal, 22*(5), 403-433.

Thompson, J. A. (2005). Proactive Personality and Job Performance: A Social Capital Perspective. *Journal of Applied Psychology, 90*(5), 1011-1017.

Tinsley, H. E. A., & Tinsley, D. J. (1987). Uses of Factor Analysis in Counseling Psychology Research. *Journal of Counseling Psychology, 34*(4), 414-424.

Tiwana, A. (2008). Do Bridging Ties Complement Strong Ties? An Empirical Examination of Alliance Ambidexterity. *Strategic Management Journal, 29*(3), 251-272.

Tortoriello, M., & Krackhardt, D. (2010). Activating Cross-Boundary Knowledge: The Role of Simmelian Ties in the Generation of Innovations. *Academy of Management Journal, 53*(1), 167-181.

Tortoriello, M., Reagans, R., & McEvily, B. (2012). Bridging the Knowledge Gap: The Influence of Strong Ties, Network Cohesion, and Network Range on the Transfer of Knowledge Between Organizational Units. *Organization Science, 23*(4), 1024-1039.

Tran, Y., Hsuan, J., & Mahnke, V. (2011). How Do Innovation Intermediaries Add Value? Insight from New Product Development in Fashion Markets. *R&D Management, 41*(1), 80-91.

Trott, P., & Hartmann, D. (2009). Why "Open Innovation" Is Old Wine in New Bottles. *International Journal of Innovation Management, 13*(4), 715-736.

Tsai, W. (2000). Social Capital, Strategic Relatedness and the Formation of Intraorganizational Linkages. *Strategic Management Journal, 21*(9), 925-939.

Tsai, W. (2001). Knowledge Transfer in Intraorganizational Networks: Effects of Network Position and Absorptive Capacity on Business Unit Innovation and Performance. *Academy of Management Journal, 44*(5), 996-1004.

Tsai, W. (2002). Social Structure of "Coopetition" within a Multiunit Organization: Coordination, Competition, and Intraorganizational Knowledge Sharing. *Organization Science, 13*(2), 179-190.

Tsai, W., & Ghoshal, S. (1998). Social Capital and Value Creation: The Role of Intrafirm Networks. *Academy of Management Journal, 41*(4), 464-476.

Tupes, E. C., & Christal, R. E. (1961). Recurrent Personality Factors Based on Trait Ratings. Lackland Air Force Base, TX: Aeronautical Systems Division, Personnel Laboratory.

Un, C. A., Cuervo-Cazurra, A., & Asakawa, K. (2010). R&D Collaborations and Product Innovation. *Journal of Product Innovation Management, 27*(5), 673-689.

Uzzi, B. (1997). Social Structure and Competition in Interfirm Networks: The Paradox of Embeddedness. *Administrative Science Quarterly, 42*(1), 35-67.

Uzzi, B. (1999). Embeddedness in the Making of Financial Capital: How Social Relations and Networks Benefit Firms Seeking Financing. *American Sociological Review, 64*(4), 481-505.

Vandermey, A. (2012). Wheels of Tomorrow. *Fortune.*

Verworn, B. (2009). A Structural Equation Model of the Impact of the "Fuzzy Front End" on the Success of New Product Development. *Research Policy, 38*(10), 1571-1581.

Villani, K. E. A. (1975). Personality/Life Style and Television Viewing Behavior. *Journal of Marketing Research, 7*(11), 432-440.

Vissa, B. (2011). A Matching Theory of Entrepreneurs' Tie Formation Intentions and Initiation of Economic Exchange. *Academy of Management Journal, 54*(1), 137-158.

Vissa, B., & Chacar, A. S. (2009). Leveraging Ties: The Contingent Value of Entrepreneurial Teams' External Advice Networks on Indian Software Venture Performance. *Strategic Management Journal, 30*(11), 1179-1191.

Walker, G., Kogut, B., & Shan, W. (1997). Social Capital, Structural Holes and the Formation of an Industry Network. *Organization Science, 8*(2), 109-125.

Wanberg, C. R., Kanfer, R., & Banas, J. T. (2000). Predictors and Outcomes of Networking Intensity Among Unemployed Job Seekers. *Journal of Applied Psychology, 85*(4), 491-503.

Weiber, R., & Mühlhaus, D. (2010). *Strukturgleichungmodellierung*. Heidelberg: Springer.

Weintraub, A. (2002). High Tech's Future is in the Toy Chest. *Business Week*, pp. 124-126.

Wenger, E. C., & Snyder, W. M. (2000). Communities of Practice: The Organizational Frontier. *Harvard Business Review, 78*(1), 139-145.

West, J., & Lakhani, K. R. (2008). Getting Clear About Communities in Open Innovation. *Industry and Innovation, 15*(2), 223-231.

Whittington, K. B., Owen-Smith, J., & Powell, W. W. (2009). Networks, Propinquity, and Innovation in Knowledge-Intensive Industries. *Administrative Science Quarterly, 54*(1), 90-122.

Wiggins, R. R., & Ruefli, T. W. (2005). Schumpeter's Ghost: Is Hypercompetition Making the Best of Times Shorter? *Strategic Management Journal, 26*(10), 887-911.

Wolff, H.-G., & Kim, S. (2012). The Relationship Between Networking Behaviors and the Big Five Personality Dimensions. *Career Development International, 17*(1), 43-66.

Wong, S.-S., & Boh, W. F. (2010). Leveraging the Ties of Others to Build a Reputation for Trustworthiness Among Peers. *Academy of Management Journal, 53*(1), 129-148.

Xiao, Z., & Tsui, A. S. (2007). When Brokers May Not Work: The Cultural Contingency of Social High-Tech Firms. *Administrative Science Quarterly, 52*(1), 1-31.

Yin, R. K. (1981). The Case Study Crisis: Some Answers. *Administrative Science Quarterly, 26*(1), 58-65.

Yin, R. K. (2009). *Case Study Research*. Thousand Oaks: Sage Publications.

Yli-Renko, H., Autio, E., & Sapienza, H. J. (2001). Social Capital, Knowledge Acquisition, and Knowledge Exploitation in Young Technology-Based Firms. *Strategic Management Journal, 22*, 587-613.

Zaheer, A., & Soda, G. (2009). Network Evolution: The Origins of Structural Holes. *Administrative Science Quarterly, 54*(1), 1-31.

Zinnbauer, M., & Eberl, M. (2004). Die Überprüfung von Spezifikation und Güte von Strukturgleichungsmodellen: Verfahren und Anwendung.

Zucker, L. G., Darby, M. R., & Armstrong, J. S. (2002). Commercializing Knowledge: University Science, Knowledge Capture, and Firm Performance in Biotechnology. *Management Science, 48*(1), 138-153.

de Jong, B. A., & Elfring, T. (2010). How Does Trust Affect The Performance of Ongoing Teams? The Mediating Role of Reflexivity, Monitoring, and Effort. *Academy of Management Journal, 53*(3), 535-549.

de Luca, L. M., & Atuahene-Gima, K. (2007). Market Knowledge Dimensions and Cross-Functional Collaboration: Examining the Different Routes to Product Innovation Performance. *Journal of Marketing, 71*(1), 95-112.

di Gangi, P. M., & Wasko, M. (2009). Steal My Idea! Organizational Adoption of User Innovations from a User Innovation Community: A Case Study of Dell IdeaStorm. *Decision Support Systems, 48*(1), 303-312.

di Gangi, P. M., Wasko, M. M., & Hooker, R. E. (2010). Getting Customers' Ideas to Work for You: Learning from Dell How to Succeed with Online User Innovation Communities. *MIS Quarterly Executive, 9*(4), 213-228.

van de Ven, A. H. (1986). Central Problems in the Management of Innovation. *Management Science, 32*(5), 590-607.

von Hippel, C., Mangum, S. L., Greenberger, D. B., Henemann, R. L., & Skoglind, J. D. (1997). Temporary Employment: Can Organizations and Employees Both Win? *Academy of Management Executive, 11*(1), 93-104.

von Hippel, E. (1986). Lead Users: A Source of Novel Product Concepts. *Management Science, 32*(7), 791-805.

von Hippel, E. (1994). "Sticky Information" and the Locus of Problem Solving: Implications for Innovation. *Management Science, 40*(4), 429-439.

von Hippel, E. (2001). Innovation by User Communities: Learning from Open-Source Software. *MIT Sloan Management Review, 42*(4), 82-86.

von Hippel, E., & von Krogh, G. (2003). Open Source Software and the "Private-Collective" Innovation Model: Issues for Organization Science. *Organization Science, 14*(2), 209-223.

Annexes

Annex 1: Related publications and data collection

I have cowritten a number of articles related to the studies presented in this dissertation. Most closely related is the following article presented at the International Product Development Management Conference 2012 in Manchester[48]:

Dumbach, M., Danzinger, F., & Möslein, K. M. (2012). Rewiring Employees Through Innovation Communities – Investigating the Antecedents of Social Capital. International Product Development Management Conference. Manchester.

The paper is coauthored by Frank Danzinger and Kathrin M. Möslein and looks at antecedents of social capital in corporate innovation communities. The article presents the findings of the pilot case of the multiple case study presented in part III. Hence, the paper can be considered as an earlier version of the research discussed in this dissertation. Some parts of the paper remained unchanged and verbatim. As the first author, I was mainly responsible for data gathering and analyzing as well as writing the paper. I received valuable inputs from the two coauthors.

Other articles I have coauthored that are based on the same data or the same theoretical background but deal with different research questions are:

Danzinger, F., & Dumbach, M. (2011). Communities for Innovation as Enablers of Cyclical Ambidexterity in SMEs. Tagungsband der Informatik 2011.

Danzinger, F., Dumbach, M., & Möslein, K. M. (2012). At the Edge and Innovative – The Role of Boundary Innovators in Demographic Change. European Academy of Management Conference. Rotterdam.

Rass, M., Dumbach, M., Danzinger, F., Bullinger, A. C., & Möslein, K. M. (2011). Open Innovation and Firm Performance: Investigating the Mediating Role of Social Capital. R&D Management Conference. Norrköping.

Concerning data collection, the qualitative data analyzed in part III was gathered in the course of three research projects at the Chair of Information Systems - Innovation and Value Creation at FAU. Researchers from the Chair of Information Systems -

[48] Unless the papers mentioned here are cited within this dissertation, they are not listed in the references.

Innovation and Value Creation as well as from other research institutes were involved in the design of the interview guidelines as well as in conducting the interviews. The qualitative data analyzed in part III is also analyzed in studies in which I am not involved. These studies focus on different research questions.

In the quantitative study (part IV), I collaborated in the process of data collection with Matthias Rass and Lucas Calmbach. They both are planning to use parts of the data set in their dissertation projects for their respective research questions. I adapted the wording of the items for the constructs representing the "Big 5" as well as intrinsic and extrinsic motivations. The wording of the scales for the social capital constructs has been adapted by Rass and myself in an intense collaborative process. The contribution from both researchers in this context was equal. Calmbach developed the wording of the items for the online and offline interaction channels in the WIN Innovation Community and approved the usage in this dissertation.

Annex 2: German wording of items used in part IV

Table 44: German wording of items (I)

STR_SUM	Nennen Sie Personen außerhalb Ihres Teams, mit denen Sie bezüglich des WIN Contest[49] in Kontakt gestanden haben. Sie können die Namen von bis zu 5 externen Kontakten angeben.
TRU_1_STD	Welchen der genannten Personen außerhalb Ihres Teams können Sie vertrauen, ohne Bedenken zu haben, ausgenutzt zu werden?
TRU_2_STD	Welche der genannten Personen außerhalb ihres Teams werden immer ihre Versprechen gegenüber Ihnen halten?
SW_1_STD	Wie regelmäßig haben Sie mit den genannten Personen außerhalb Ihres Teams bezüglich der Aufgabe im WIN Contest in Kontakt gestanden?
SW_2_STD	Wie regelmäßig haben Sie sich mit den genannten Personen außerhalb Ihres Teams gegenseitig unterstützt?
SW_3_STD	Wie eng waren die Arbeitsbeziehungen mit den genannten Personen außerhalb Ihres Teams?
SW_4_STD	Wie würden Sie die Unterstützung mit den genannten Personen außerhalb Ihres Teams beschreiben?
	Meine Kontakte außerhalb meines Teams und ich…
CGN_1	…hatten dieselben Ambitionen und dieselbe Vision.
CGN_2	…verfolgten begeistert gemeinsame Ziele und eine gemeinsame Mission.
CGN_3	…verwendeten stets dieselbe Sprache und dieselben Begrifflichkeiten.
CGN_4	…haben uns sprachlich problemlos verstanden.

[49] As I have already mentioned, the WIN Innovation Community also comprises aspects of an innovation contest (e.g., Adamczyk et al., 2011). As community members were more familiar with the term "WIN Contest", I used this term in the German wording of the items.

Table 45: German wording of items (II)

	Mir ist es wichtig...
MOT_EX_1	...eine gute Note in der Veranstaltung Grundlagen des E-Business zu erreichen.
MOT_EX_2	...eine gute Abschlussnote im Bachelor zu erhalten.
MOT_EX_3	...eine überdurchschnittliche gute Note in der Veranstaltung Grundlagen des E-Business zu erreichen.
MOT_EX_4	...eine gute Note in der Veranstaltung Grundlagen des E-Business zu erreichen, damit ich anderen meine Fähigkeiten unter Beweis stellen kann (z.B. meiner Familie oder potenziellen Arbeitgebern).
MOT_INT_1	Ich habe Freude daran am WIN Contest teilzunehmen.
MOT_INT_2	Ich arbeite gerne für den WIN Contest, um gemeinsam mit anderen Ideen zu entwickeln und zu verbessern.
MOT_INT_3	Es gibt mir ein gutes Gefühl, anderen bei der Entwicklung von Ideen im WIN Contest zu helfen.
MOT_INT_4	Für mich ist es von größter Bedeutung, Spaß an der Teilnahme am WIN Contest zu haben.
	Bitte beschreiben Sie sich anhand der untenstehenden Liste von Eigenschaften selbst. Beschreiben Sie sich bitte so, wie Sie momentan sind, nicht wie Sie zukünftig sein möchten. Beschreiben Sie sich bitte im Vergleich zu anderen Personen des gleichen Geschlechts und eines ungefähr gleichen Alters.
EST_1	Launisch
EST_2	Temperamentvoll
EST_3	Leicht reizbar
INT_1	Angespannt im Umgang mit anderen Personen
INT_2	Einzelgänger
INT_3	Schüchtern
OTE_1	Kreativ
OTE_2	Phantasievoll
OTE_3	Einfallsreich
AGR_1	Verständnisvoll
AGR_2	Freundlich
AGR_3	Warmherzig
CON_1	Gewissenhaft
CON_2	Sorgfältig
CON_3	Gut organisiert

Table 46: German wording of items (III)

	Ich bin davon überzeugt, dass...
TRU_B_1	...ich mich bei Problemen auf die Hilfe der genannten Personen außerhalb meines Teams verlassen kann.
TRU_B_2	...die genannten Personen außerhalb meines Teams meine Interessen bei wichtigen Entscheidungen mit einbeziehen.
TRU_B_3	...mich die genannten Personen außerhalb meines Teams über wichtige Angelegenheiten auf dem Laufenden halten.
TRU_B_4	...ich mich darauf verlassen kann, dass die genannten Personen außerhalb meines Teams zu ihrem Wort stehen.
TRU_B_5	...ich den genannten Personen außerhalb meines Teams vertrauen kann.

Annex 3: Evaluation of the measurement of trust in part IV

In order to evaluate the measurement model validity of trust based on the name generator further, an additional approach was taken. While I decided to benefit from the strengths of the name generator method, trust was measured in a second way which was not based on the name generator. The construct was measured with five indicators adopted from de Jong and Elfring (2010), while I deleted two because of insufficient parsimony (RMSEA of 0.172 and AGFI of 0.830 for the five indicator measurement, and RMSEA of 0.139 for the four indicator measurement). Table 47 summarizes the results of the additional measurement of trust. The exact wording of the items was (1) "I am able to count on my team-external contacts for help if I have difficulties", (2) "I am confident that my team-external contacts will take my interests into account when making community-related decisions", (3) "I am confident that my team-external contacts will keep me informed about issues that concern community work", (4) "I can rely on my team-external contacts to keep their word", and (5) "I trust my team-external contacts". As a next step, I estimated the correlation between the construct of trust measured on the basis of the name generator as well as the additional trust construct. The results suggest that the model fits the data well (SRMR=0.0028; RMSEA=0.039; CFI=0.999; GFI=0.996; AGFI=0.985). In addition, the model suggests a correlation coefficient of 0.74, which indicates that both constructs represent similar phenomena. I conclude that my analysis indicates valid measurement of trust.

Table 47: Measurement validity of the alternative measurement of trust

SC_TRU_B: Social capital: Trust (B)				
Description of Indicator	*Mean (standard deviation)*	*Item-to-total correlation*	*Item reliability*	*t-value*
SC_TRU_B_1: "I am able to count on my team-external contacts for help if I have difficulties"	4.4270 (2.52107)	0.966	0.949	-
SC_TRU_B_2: "I am confident that my team-external contacts will take my interests into account when making community-related decisions"	4.0967 (2.38162)	Deleted based on RMSEA and AGFI		
SC_TRU_B_3: "I am confident that my team-external contacts will keep me informed about issues that concern community work"	4.2860 (2.46278)	Deleted based on RMSEA and AGFI		
SC_TRU_B_4: "I can rely on my team-external contacts to keep their word"	4.4532 (2.50606)	0.971	0.962	101.547***
SC_TRU_B_5: "I trust my team-external contacts"	4.5287 (2.54663)	0.974	0.971	106.655***
First generation indices		**Second generation indices (local)**		
Cronbach's alpha	0.987	Comp. reliability	0.99	
		AVE	0.96	
		F-L-ratio	0.94	

Annex 4: Covariance matrix of items used in part IV

Table 48: Covariance matrix (I)

Items	STR_SUM	TRU_1_STD	TRU_2_STD	SW_1_STD	SW_2_STD	SW_3_STD	SW_4_STD	CGN_1	CGN_2	CGN_3	CGN_4
STR_SUM	2.273	.351	.328	.242	.197	.229	.249	1.963	1.863	2.430	2.663
TRU_1_STD	.351	.070	.063	.039	.032	.038	.042	.347	.319	.423	.459
TRU_2_STD	.328	.063	.066	.037	.031	.036	.040	.326	.304	.405	.435
SW_1_STD	.242	.039	.037	.034	.027	.029	.031	.225	.217	.271	.289
SW_2_STD	.197	.032	.031	.027	.029	.027	.028	.205	.202	.226	.241
SW_3_STD	.229	.038	.036	.029	.027	.033	.032	.240	.236	.269	.285
SW_4_STD	.249	.042	.040	.031	.028	.032	.036	.258	.249	.291	.311
CGN_1	1.963	.347	.326	.225	.205	.240	.258	4.540	4.048	4.246	4.385
CGN_2	1.863	.319	.304	.217	.202	.236	.249	4.048	4.370	4.050	4.129
CGN_3	2.430	.423	.405	.271	.226	.269	.291	4.246	4.050	5.979	5.950
CGN_4	2.663	.459	.435	.289	.241	.285	.311	4.385	4.129	5.950	6.678
MOT_EX_1	.134	.023	.025	.018	.017	.016	.019	.226	.215	.303	.312
MOT_EX_2	.129	.026	.026	.018	.016	.019	.020	.249	.234	.366	.351
MOT_EX_3	.063	.014	.019	.009	.017	.012	.013	.176	.187	.163	.193
MOT_EX_4	.126	.021	.027	.019	.031	.026	.023	.206	.340	.226	.230
MOT_INT_1	.187	.031	.031	.029	.030	.026	.026	.353	.396	.381	.322
MOT_INT_2	.199	.034	.031	.034	.034	.031	.029	.444	.535	.446	.403
MOT_INT_3	.223	.046	.047	.037	.040	.040	.036	.460	.520	.482	.457
MOT_INT_4	.194	.036	.036	.038	.034	.036	.034	.405	.463	.506	.451
EST_1	.004	.008	.009	.007	.002	.000	.002	-.044	-.027	-.093	-.108
EST_2	.168	.019	.018	.027	.027	.025	.024	.230	.237	.215	.243
EST_3	.020	.005	.013	.006	.002	.000	.001	.059	.029	-.044	-.066
EXT_1	-.183	-.029	-.035	-.027	-.030	-.029	-.030	-.237	-.215	-.268	-.303
EXT_2	-.273	-.055	-.057	-.045	-.038	-.044	-.045	-.426	-.434	-.485	-.524
EXT_3	-.120	-.010	-.014	-.023	-.028	-.022	-.019	-.168	-.182	-.157	-.193
OTE_1	.079	.017	.018	.014	.017	.017	.014	.228	.262	.133	.180
OTE_2	.112	.023	.027	.018	.021	.025	.023	.219	.265	.140	.211
OTE_3	.073	.015	.018	.012	.018	.019	.018	.198	.205	.102	.151
AGR_1	.090	.020	.026	.013	.014	.020	.020	.190	.210	.238	.249
AGR_2	.122	.025	.026	.014	.013	.014	.016	.171	.158	.204	.248
AGR_3	.181	.032	.038	.025	.023	.030	.029	.405	.445	.452	.431
CON_1	.156	.027	.034	.016	.017	.021	.024	.288	.346	.264	.297
CON_2	.110	.017	.030	.016	.025	.024	.024	.338	.400	.314	.288
CON_3	.080	.007	.013	.014	.018	.019	.015	.300	.374	.289	.249

Table 49: Covariance matrix (II)

Items	MOT_EX_1	MOT_EX_2	MOT_EX_3	MOT_EX_4	MOT_INT_1	MOT_INT_2	MOT_INT_3	MOT_INT_4	EST_1	EST_2	EST_3
STR_SUM	.134	.129	.063	.126	.187	.199	.223	.194	.004	.168	.020
TRU_1_STD	.023	.026	.014	.021	.031	.034	.046	.036	.008	.019	.005
TRU_2_STD	.025	.026	.019	.027	.031	.031	.047	.036	.009	.018	.013
SW_1_STD	.018	.018	.009	.019	.029	.034	.037	.038	.007	.027	.006
SW_2_STD	.017	.016	.017	.031	.030	.034	.040	.034	.002	.027	.002
SW_3_STD	.016	.019	.012	.026	.026	.031	.040	.036	.000	.025	.000
SW_4_STD	.019	.020	.013	.023	.026	.029	.036	.034	.002	.024	.001
CGN_1	.226	.249	.176	.206	.353	.444	.460	.405	-.044	.230	.059
CGN_2	.215	.234	.187	.340	.396	.535	.520	.463	-.027	.237	.029
CGN_3	.303	.366	.163	.226	.381	.446	.482	.506	-.093	.215	-.044
CGN_4	.312	.351	.193	.230	.322	.403	.457	.451	-.108	.243	-.066
MOT_EX_1	.907	.534	.799	.672	.366	.383	.348	.339	.002	.088	-.010
MOT_EX_2	.534	.650	.579	.436	.200	.231	.196	.172	.001	.105	.045
MOT_EX_3	.799	.579	2.000	1.266	.515	.525	.431	.391	-.039	.111	.037
MOT_EX_4	.672	.436	1.266	3.055	.877	.832	.838	.682	.079	.323	.111
MOT_INT_1	.366	.200	.515	.877	2.697	2.302	1.917	1.567	-.103	.242	-.225
MOT_INT_2	.383	.231	.525	.832	2.302	2.680	2.088	1.692	-.161	.253	-.233
MOT_INT_3	.348	.196	.431	.838	1.917	2.088	2.629	1.657	-.183	.176	-.273
MOT_INT_4	.339	.172	.391	.682	1.567	1.692	1.657	2.702	-.083	.252	-.143
EST_1	.002	.001	-.039	.079	-.103	-.161	-.183	-.083	2.450	.947	1.349
EST_2	.088	.105	.111	.323	.242	.253	.176	.252	.947	2.349	.960
EST_3	-.010	.045	.037	.111	-.225	-.233	-.273	-.143	1.349	.960	2.411
EXT_1	-.041	-.061	-.016	.038	-.066	-.165	-.061	-.079	.518	-.051	.548
EXT_2	-.102	-.150	.029	-.225	-.126	-.141	-.181	-.190	.465	.035	.391
EXT_3	-.064	-.024	-.088	-.051	-.049	-.095	-.003	.046	.394	-.390	.227
OTE_1	.209	.148	.202	.225	.655	.699	.549	.562	-.041	.353	-.190
OTE_2	.141	.068	.136	.165	.565	.560	.484	.490	.026	.439	-.107
OTE_3	.200	.098	.257	.144	.515	.533	.489	.389	-.121	.259	-.183
AGR_1	.177	.150	.103	.162	.133	.232	.243	.248	-.237	-.084	-.321
AGR_2	.193	.138	.119	.172	.188	.220	.263	.256	-.220	.013	-.257
AGR_3	.239	.202	.129	.274	.291	.308	.361	.359	-.168	.082	-.164
CON_1	.372	.297	.472	.285	.180	.231	.237	.186	-.157	.065	-.109
CON_2	.363	.334	.501	.329	.226	.302	.268	.175	-.240	-.050	-.112
CON_3	.321	.292	.425	.335	.257	.336	.300	.316	-.274	.047	-.133

Table 50: Covariance matrix (III)

Items	EXT_1	EXT_2	EXT_3	OTE_1	OTE_2	OTE_3	AGR_1	AGR_2	AGR_3	CON_1	CON_2	CON_3
STR_SUM	-.183	-.273	-.120	.079	.112	.073	.090	.122	.181	.156	.110	.080
TRU_1_STD	-.029	-.055	-.010	.017	.023	.015	.020	.025	.032	.027	.017	.007
TRU_2_STD	-.035	-.057	-.014	.018	.027	.018	.026	.026	.038	.034	.030	.013
SW_1_STD	-.027	-.045	-.023	.014	.018	.012	.013	.014	.025	.016	.016	.014
SW_2_STD	-.030	-.038	-.028	.017	.021	.018	.014	.013	.023	.017	.025	.018
SW_3_STD	-.029	-.044	-.022	.017	.025	.019	.020	.014	.030	.021	.024	.019
SW_4_STD	-.030	-.045	-.019	.014	.023	.018	.020	.016	.029	.024	.024	.015
CGN_1	-.237	-.426	-.168	.228	.219	.198	.190	.171	.405	.288	.338	.300
CGN_2	-.215	-.434	-.182	.262	.265	.205	.210	.158	.445	.346	.400	.374
CGN_3	-.268	-.485	-.157	.133	.140	.102	.238	.204	.452	264	.314	.289
CGN_4	-.303	-.524	-.193	.180	.211	.151	.249	.248	.431	.297	.288	.249
MOT_EX_1	-.041	-.102	-.064	.209	.141	.200	.177	.193	.239	.372	.363	.321
MOT_EX_2	-.061	-.150	-.024	.148	.068	.098	.150	.138	.202	.297	.334	.292
MOT_EX_3	-.016	.029	-.088	.202	.136	.257	.103	.119	.129	.472	.501	.425
MOT_EX_4	.038	-.225	-.051	.225	.165	.144	.162	.172	.274	.285	.329	.335
MOT_INT_1	-.066	-.126	-.049	.655	.565	.515	.133	.188	.291	.180	.226	.257
MOT_INT_2	-.165	-.141	-.095	.699	.560	.533	.232	.220	.308	.231	.302	.336
MOT_INT_3	-.061	-.181	-.003	.549	.484	.489	.243	.263	.361	.237	.268	.300
MOT_INT_4	-.079	-.190	.046	.562	.490	.389	.248	.256	.359	.186	.175	.316
EST_1	.518	.465	.394	-.041	.026	-.121	-.237	-.220	-.168	-.157	-.240	-.274
EST_2	-.051	.035	-.390	.353	.439	.259	-.084	.013	.082	.065	-.050	.047
EST_3	.548	.391	.227	-.190	-.107	-.183	-.321	-.257	-.164	-.109	-.112	-.133
EXT_1	2.197	.921	1.168	-.131	-.167	-.187	-.175	-.130	-.110	-.061	-.087	-.270
EXT_2	.921	2.609	.952	-.110	-.121	-.062	-.314	-.240	-.340	-.033	-.147	-.210
EXT_3	1.168	.952	2.591	-.206	-.171	-.283	.017	-.023	.051	.040	-.002	-.169
OTE_1	-.131	-.110	-.206	2.014	1.421	1.144	.347	.300	.329	.340	.283	.243
OTE_2	-.167	-.121	-.171	1.421	1.925	1.124	.341	.289	.315	.224	.138	.062
OTE_3	-.187	-.062	-.283	1.144	1.124	1.475	.284	.311	.237	.344	.271	.235
AGR_1	-.175	-.314	.017	.347	.341	.284	1.271	.651	.794	.562	.593	.497
AGR_2	-.130	-.240	-.023	.300	.289	.311	.651	.856	.692	.491	.484	.431
AGR_3	-.110	-.340	.051	.329	.315	.237	.794	.692	1.510	.555	.510	.468
CON_1	-.061	-.033	.040	.340	.224	.344	.562	.491	.555	1.587	1.235	1.076
CON_2	-.087	-.147	-.002	.283	.138	.271	.593	.484	.510	1.235	1.751	1.339
CON_3	-.270	-.210	-.169	.243	.062	.235	.497	.431	.468	1.076	1.339	2.180

MIX
Papier aus verantwortungsvollen Quellen
Paper from responsible sources
FSC® C105338

If you have any concerns about our products,
you can contact us on
ProductSafety@springernature.com

In case Publisher is established outside the EU,
the EU authorized representative is:
Springer Nature Customer Service Center GmbH
Europaplatz 3, 69115 Heidelberg, Germany

Printed by Libri Plureos GmbH
in Hamburg, Germany